Are We Hardwired?

Also by William R. Clark

The Experimental Foundations of Modern Immunology
At War Within: The Double-Edged Sword of the Immune System
Sex and the Origins of Death
The New Healers: The Promise and Problems of Molecular Medicine in the Twenty-first Century
A Means to an End: The Biological Basis of Aging and Death

Are We Hardwired?

The Role of Genes in Human Behavior

William R. Clark

Michael Grunstein

OXFORD
UNIVERSITY PRESS
2000

OXFORD
UNIVERSITY PRESS

Oxford New York
Athens Auckland Bangkok Bogotá Buenos Aires Calcutta
Cape Town Chennai Dar es Salaam Delhi Florence Hong Kong Istanbul
Karachi Kuala Lumpur Madrid Melbourne Mexico City Mumbai
Nairobi Paris São Paulo Shanghai Singapore Taipei Tokyo Toronto Warsaw

and associated companies in
Berlin Ibadan

Published by Oxford University Press, Inc.
198 Madison Avenue, New York, NY 10016

Library of Congress Cataloging-in-Publication Data

Clark, William R., 1938—
Are we hardwired?: the role of genes in human behavior /
by William R. Clark, Michael Grunstein.
p. cm.
Includes bibliographical references and index.
ISBN 0-19-513826-0
1. Behavior genetics. I. Grunstein, Michael. II. Title.
QH457.C58 2000
155.7–dc21 99-054699

9 8 7 6 5 4 3 2 1

Printed in the United States of America
on acid free paper

Contents

Prologue vii

1. Mirror, Mirror 3
2. In the Beginning: The Evolutionary Origins of Behavior 23
3. The Nose Knows 41
4. As the Worm Turns: Learning and Memory in the Roundworm *C. elegans* 59
5. About Genes and Behavior 75
6. Life in the Fourth Dimension: The Role of Clocks in Regulating Behavior 95
7. You Must Remember This: The Evolution of Learning and Memory 117
8. The Role of Neurotransmitters in Human Behavior 137
9. The Genetics of Aggression 157
10. The Genetics of Consumption, Part I: Eating Disorders 177
11. The Genetics of Consumption, Part II: Substance Abuse 199
12. The Genetics of Human Mental Function 221
13. The Genetics of Human Sexual Preference 239
14. Genes, the Environment, and Free Will 253

Appendix I Finding and Identifying Genes 271
Appendix II A Brief History of Eugenics 279
References 295
Index 313

Prologue

Why are human beings so different from one another? Why are some people tall, some short; some brown-eyed, some blue-eyed? The fact that daughters and sons tend to look like their mothers and fathers suggests that physical features are heritable and therefore due in large part to genes. But what about behavior? Why, in the very same family, are some children assertive, others shy? Why are some people confident, others uncertain? Why are some highly emotional, others more reserved and "logical?" Are these traits heritable, too? Do differences in these traits among individuals also have a genetic basis?

The idea that at least some of the variability we see in human behavior and personality is heritable, and therefore genetically determined, would certainly come as no surprise to most animal breeders. For at least half a millennium or more, animals have been bred specifically to reinforce certain behavioral or personality traits. Some dogs, ranging in size from tiny terriers to massive pit bulls or Dobermans, have been bred for their aggressive nature. Others, such as collies or spaniels, faithfully transmit a docile, loving nature from generation to generation. Still others have been bred to carry out specific tasks related to hunting or managing flocks of animals. In the laboratory, rats and mice have been selectively bred for many generations to create strains that are fearful or aggressive. These strains pass on their personality differences each time they breed. No one seriously questions the role of genes in the development of animal behavior, or of inheritance in passing these traits from one generation to the next. Yet we are reluctant to acknowledge a similar role of genes in guiding human behavior.

At a deeper level, we know that the lives of cells are closely governed by genes, whether those cells are individual, free-living organisms such as yeast and amoebas, or the interactive cells that make up our own bodies. Single-cell organisms show definite signs of behavior. Within ourselves, the cells charged with managing how we react, how we behave, are all located within our nervous systems. And it is precisely when we come to the cellularly more complex nervous system that the issue of genes and behavior becomes complicated. The major complication arises because, more than in any other organ or system in the body, the behavior of cells in the nervous system is affected not only by genes but by the external environment. Nerve cells are our window onto the world around us. We use the various images impressed onto our nerve cells to formulate responses to our environment, and this experience of our environment—and our responses to it—are remembered. Nerve cells are altered by contact with the environment, in ways that are still only crudely understood but which alter the way we respond to the same information when it appears in the environment again.

The role of genes in governing behavior remains one of the most controversial topics in all of human biology. Early in this century, overeager promotion of a genetic basis for behavior led to the initial silly excesses but ultimately to the stunning horrors of eugenics. Subsequent reactions to these excesses, within both the scientific community and society as a whole, led to a nearly complete dismissal of a role for genes in human behavior for many decades. We are slowly coming back to a more balanced view. A detailed study of the biological basis of behavior in animals, from the simplest single-celled creatures through the most complex mammals, shows that genes play a very important role in guiding behavior. Inheritance studies in humans, especially those involving twins reared together or apart, indicate clearly that humans are no exception. The variability we see around us in the way humans respond in a given situation is to a large extent influenced by the variability in their genetic makeup.

Part of our concern about the role of genes in determining human behavior surely lies in our concerns about free will and personal responsibility. All legal and moral systems assume that individuals are free to choose among alternative courses of behavior; individual responsibility has no meaning in the absence of unimpeded choice. But if our every behavior can be predicted from what is written into our genes before we are born, what does that say about our freedom to choose, or about individual responsibility for the choices we make? The opposite view, that we come into the world as some sort of blank template and that what we become, as cognizant adults, is simply the totality of our previous experiences, renders our behavior no less predictable—and our freedom to choose and act no less constrained—than if we were simply the sum of our genes.

As we begin the new millennium, we will witness the completion of one of the boldest undertakings in the history of biology and medicine—the Human Genome Project. This project, when completed in 2003, will provide us with a complete catalog of every gene involved in the construction and operation of an entire human being. It will take another decade or two to sort out what most of these genes do, but beyond question many of them will be involved in fundamental ways in determining human behavior. Understanding the extent to which certain behaviors are genetically determined will certainly provide us with better insight into human nature and give us a more realistic sense of the extent to which behavior can be modified by external interventions. That in itself will be of enormous value. But we will also have at our command the technology to alter those genes. What will we do with this new information? How will we explore these new possibilities? Will it be possible to use this information for the betterment of humankind, or will we simply plunge headlong into a renewed and even more disturbing flirtation with eugenics?

These are important questions, much too important to leave to scientists and politicians. All of us must become involved with these issues, and we can become involved only if we understand not just the questions themselves, but the information and technology from which these questions emerge. That is the purpose of this book. We will explore behavior at its most fundamental level, beginning with the lives of individual cells. We will build upon the knowledge we gain at this level and follow the evolutionary pathways leading to human behavior. We will find that there is in fact little that is new in human behavior, when it is analyzed in terms of underlying molecular and genetic mechanisms—most were already in place billions of years ago. And, most important, it doesn't take a Ph.D. in biology to understand them. It is well within anyone's grasp to learn a great deal about one of the most important and compelling issues of our time—the biological basis of human behavior.

Are We Hardwired?

1

Mirror, Mirror

In August 1939, in the tiny town of Piqua in northeast Ohio, an unmarried woman gave birth, slightly prematurely, to twin boys. The country as a whole was struggling through the last of the depression years, and times were still hard in Ohio, as in most other places in prewar America. This poor mother, a recent immigrant, simply had no way of supporting two newborn infants, so, like many others who found themselves in her position, she asked the hospital to help her place the boys for adoption. These were healthy, handsome babies, and they were taken quickly; both were adopted within a few weeks, both by middle-class working families. One of the families was from Piqua itself; the other was from across the state, in Lima. In what would be the first of a string of improbable coincidences involving these twins, both families named their newly adopted son James. But aside from these first few weeks in the same hospital, the two "Jimmies" would remain largely unaware of each other's existence, and they would not be brought together again for nearly forty years. Only then would they learn that they were genetically identical—that they were, in a sense, clones of one another.

Just a few years later, in a small city in western Michigan, another set of twin boys was born, again to an immigrant mother, but this time to one who was fortunate enough to have a husband and, even more fortunate, a husband with a job. It didn't pay much—he stitched and delivered burlap sacks to potato growers in the area—but it was enough to keep them all together. The twins came at the end of a string of four other children, and were seen as a mixed blessing for an already hard-strapped family. But they were welcomed; the family just squeezed a bit more tightly together in their modest house. Unlike the two Jimmies, these two boys would grow up in the same home, with exactly the same siblings and neighbors and pets, for nearly eighteen years. And the Michigan twins were different in another way: They were not identical, but fraternal.

These two sets of twins would never meet; they were never, in fact, aware of each other's existence. But they present those who study the development of human behavior with a natural experiment, an experiment of the type that could never be done with human subjects in a research laboratory. They have the potential to provide us with important insights into the role of genetics versus environment, of "nature versus nurture," in shaping who we turn out to be in life. The two fraternal twins, Tommy and Bernie, were genetically distinct, yet reared in the same environment. The two Jimmies, although genetically identical, were raised in different environments, with different parents, siblings, neighbors, and pets.

One of the two Jimmies became aware through his adoptive mother that he had a twin brother, and eventually set out to find him. He was successful, and when the two were finally reunited at age thirty-nine, and began to swap childhood stories with one another, they were astounded at the number of similarities in their lives. Some were trivial, and almost certainly nothing more than monumental coincidences. For example, each had adoptive siblings named Larry and pet dogs named Toy. Both preferred Miller Lite beer, and smoked Salem cigarettes. Each had married a woman named Linda, divorced, and then married a woman named Betty. One named his firstborn son James Alan, the other James Allen. Each loved stock car racing and hated baseball.

But as time went on, they would find additional, and from a biological point of view more intriguing, similarities in their separate early lives. Their personalities, as described by family members and as clearly indicated by standardized personality assessment tests, were remarkably similar. Each had developed sinus headaches at about age ten, a condition that eventually developed into recurring migraine headaches. Their descriptions of their symptoms, elicited by specialists, were virtually identical. Each had been good at math in school, while each struggled with English; their overall scholastic performances were remarkably similar. Each had a fondness for woodworking, and each developed the habit of biting his fingernails.

Tommy and Bernie, on the other hand, were as different as they could be from the very first day. Tommy was a peaceful baby, slept a great deal, and responded positively to being held and touched. Bernie was the opposite. He fussed or cried constantly; cuddling and caressing by family members didn't seem to console him. In a matter of just a few weeks, everyone began speaking of the twins in quite different terms. Tommy, by common consent, was "an angel"; Bernie was "difficult." Before long, it was obvious that the family interacted rather differently with each of the boys. The differences in their behavior continued to develop as the boys became toddlers and started interacting with other children in the neighborhood. Tommy seemed content to stay in the house and play with his toys; Bernie charged outside, where he would engage in play with other children. He seemed to have a penchant for rough-housing, and often came home bruised or cut. These differences carried through to the boys' early years at school. Tommy was adored by all his teachers; Bernie was sent home frequently because of his disruptive behavior in class.

As the two Jimmies moved into their young adult years, during which they had no contact, they continued developing similar habits and personality traits. They expressed themselves alike, and used similar slang phrases. Both suddenly put on about ten extra pounds at roughly the same point in life. In the years preceding their ultimate reunion, both encountered assorted stresses that led to chest pains and high blood pressure. Both had difficulty sleeping, and both were taking Valium for general nervousness. Both held clerical jobs, and each had developed a fascination with police work; each had become a volunteer sheriff's assistant in his respective community.

Tommy's and Bernie's lives took very different courses. Tommy was a middling student, but went on for two additional years after high school at the local community college. After a short stint in the military, part of which was spent in Vietnam, he entered a Catholic seminary and became a priest. Bernie dropped out of high school, got into a series of scrapes with the law, and has been in and out of prison a good deal of his adult life. His social interactions have remained extremely difficult, and he never married. In spite of their extraordinarily different personalities, they have stayed in close touch over the years.

The behavior of these fraternal twins highlights a subtle but very important point about the interaction of individuals with their environment. We commonly think of the impact of the environment on the individual, but for human beings this interaction is actually a two-way street. Although Tommy and Bernie grew up in exactly the same environment, they manipulated that environment differently. This was not necessarily a willful or even conscious manipulation on their part. These were two genetically quite different individuals, and some of their differences were expressed as personality differences that were immediately perceived by the people around them. Based in turn on their own personalities, those people responded in different ways to Tommy and Bernie. It was not the environment that made one quarrelsome, and the other mild-mannered. But their differences elicited markedly different responses from those around them, and in that sense they cannot be said to have grown up in exactly the same environment.

We will talk a good deal about "environment" as we proceed through this book, because it is impossible to talk about the role of genes in behavior without taking the environment into account as well. So let us take just a moment to think about what we mean when we refer to the environment. For most animals, the environment means what we might call the physical or ecological environment, the natural surroundings in which the individual "behaves": competes for resources to stay alive, to find a mate, and to produce offspring. These basic behaviors are the same for humans as for animals, but we pursue them in two quite different sorts of environments. We, too, function within the context of an ecological environment; we, too, must eat, stay warm, find food and a mate. But unlike animals, we also function within the context of a cultural environment. Culture consists of a wide range of abstract ideas, social customs, rituals, creative works, and institutions that are largely made possible by language. The cultural environment shapes human beings every bit as forcefully as the ecological environment; in fact, it can be argued that culture is now a more important factor in human genetic evolution than is the ecological environment. So as we proceed through the chapters that follow, we should always bear in mind not only the role of environment in determining behavior in a general sense, but also the unique role of the cultural environment in determining human behavior in particular. The interaction between our genetic selves and our cultural selves is very complex indeed.

The Biological Nature of Twins

Twins have always fascinated us. They are a mystery, and so it is perhaps not surprising that they had become bound up in mythology long before they became an object of medical investigation. They are feared in some cultures, and revered in others. Twinning is much more common than live births would suggest. Although only about four sets of fraternal twins, and one set of identical twins, are produced per 1,000 live births in the United States, sophisticated sonographic studies of early human pregnancies suggest that as many as one in eight conceptions—and possibly more—produces multiple embryos. The vast majority of these are lost in the first few weeks of pregnancy, and under normal circumstances their existence is undetected by either the mother or her physician.

The most intriguing twins—identical twins—are created when a single, fertilized egg splits into two parts shortly after fertilization. This can occur at several different stages in development; the later it occurs, the more similar the twins. Once fertilization has taken place, the egg starts dividing and begins its journey through the Fallopian tube toward the uterus. The immediate product of the union of a sperm and egg is called a zygote, and so twins arising from a single fertilization event are also called monozygotic twins. Once the zygote starts dividing, the developing cell mass is called an embryo; when the embryo is clearly recognizable as human (after about five weeks), it is referred to as a fetus.

Amazingly, for a number of rounds of cell division after an egg is fertilized, entire individuals can be formed from only a portion of the cells in a developing embryo. None of the cells in the evolving cell mass has yet become specialized, and all retain the potential to produce an entire individual. Physical splitting of the cell mass, or "partitioning" as it is sometimes called, can occur at various stages in the early stages of the gestational process. This part of the twinning process is not well understood. It is not at all clear why such a coherent cell mass would break apart, or, once it did, why the parts would not simply rejoin, since embryonic cells are very sticky. Early mouse embryo cell masses can be readily separated in the laboratory, but great care has to be taken to prevent the separated parts from rejoining. The stable partitioning of an early embryonic cell mass is such an unlikely event that many obstetricians wonder whether it might be a type of birth defect, albeit an entirely harmless one.

Very early partitioning (up to three or four days post-fertilization) of human embryos results in identical twins that have separate placentas. Twins arising from partitioning events occurring after about four days (roughly 70% of monozygotic twin pairs) usually share a common placenta. There is some evidence that identical twins attached to separate placentas may develop slightly differently, and there has been a lengthy—and largely unresolved—debate about how this might affect the future development of the individuals involved. On occasion, partitioning may occur up to as late as ten days post-conception, and in these cases separation may be only partially complete, leading to various degrees of the condition known as Siamese twins. All three types of twins—shared placenta, individual placentas, and Siamese—are monozygotic and genetically identical, but it is conceivable that their different developmental patterns may result in subtle differences between the twins as adults.

More than two genetically identical individuals may arise from a single fertilization event, although this is quite rare. The developing embryonic cell mass may split into three or even more parts, resulting in multiple monozygotic individuals. The Dionne quintuplets, born in Canada in 1934, were genetically identical and thus monozygotic. It is also possible to have mixed multiple births; it is not at all unusual for quadruplets to consist of one pair of identical twins and one pair of fraternal twins, for example.

We must introduce a note of caution here about use of the term "genetically identical" with regard to monozygotic twins. Although monozygotic twins start out life from a common genetic blueprint, the development of a fully grown adult from the early zygotic stage is not a perfectly controlled process. Potentially mutagenic errors are regularly detected and removed in the line of cells giving rise to sperm or eggs in each generation, but this process is considerably less strict in production of those cells that form the soma (all of the rest of the cells in the body) during the development of a single individual. Normally, mutations that arise in somatic cells do not spread far in the body, because most somatic cells give rise to only a few progeny in their lifetimes, particularly in behavior-generating tissues such as the brain. Nevertheless, the accumulation of somatic errors throughout life is a potential source of genetic differences in otherwise identical twins.

Even in the development of two individuals from completely identical genetic blueprints, the developmental pathways might not be exactly the same. Particularly in development of the nervous system, which is at the center of all behavior, there is a good deal of randomness in the generation of varying portions of the brain and peripheral nerves. During embryonic and fetal life, newly forming nerve cells toss out fibers more or less randomly into their immediate vicinity. These will by chance form connections with other nerve cells or with nearby muscle cells. Nerve cells failing to make a connection die off; those that establish a connection retain that connection essentially for life. But even two genetically identical twins will develop slightly different patterns of nerve cell connections, and these differences could well be a basis for differences between monozygotic twins. Detailed analyses of the brains of monozygotic twins have in fact revealed small but potentially important neuroanatomical variations.

Unlike identical twins, fraternal twins are conceived when two separate sperm fertilize two separate eggs, and they are thus referred to as dizygotic twins. They never share the same placenta. The resulting embryos share the same womb at the same time, but they are no more alike genetically than any two children born to the same parents through different birth events. That makes them far more alike genetically (50% alike, on average) than two children selected from the population at random, but still rather far from complete genetic identity. Identical twins (with extremely rare exceptions) are always of the same sex; individual members of fraternal twin pairs each has a random chance of being male or female.

Until well into the present century, same-sex twins were judged to be identical or fraternal largely on the basis of appearance. Occasionally, fraternal twins may seem so alike that it would be easy to mistake them for identical twins. On rare occasions, identical twins may seem slightly different in appearance. Unless twins are of the opposite sex, they are now routinely tested for blood group proteins or DNA markers that allow doctors and parents to make an unambiguous determination of their genetic status.

What Twins Can Tell Us about the Role of Genes in Human Personality

Scientists interested in the genetics of human behavior are interested first and foremost in variability. The question really is not whether genes underlie human behavior. Ultimately every aspect of the existence of every biological organism is determined by its genes; humans are no different in this respect. The real question is to what extent variability in the genes affecting behavior contribute to the variability in human behavior we see around us, and to what extent this variability is determined by differences in the environment—the home in which the individual was raised, churches and schools attended, and the community in which the individual lives and works throughout his or her life.

The study of such questions in humans is accompanied by a number of restraints. In subsequent chapters, we look at the role of genes in causing variations in behavior in a wide range of animal species, from the simplest single-cell organisms, to fruit flies and roundworms and to mammals such as rats and mice. Our information comes from a variety of experiments that are simply not possible in humans. Laboratory animals can be selectively bred to reveal inheritance patterns from generation to generation. In many of these species, dozens of generations can be produced in a single year; humans require over a dozen years to produce a single generation. If we suspect that a particular form of a given gene causes a particular behavioral variation in an animal species, we can very often insert that gene variant into one of them to see just what it does.

We cannot do any of these things in humans. We can only observe from the outside whatever it is that humans do naturally and of their own free will. Thus one of the oldest (and still used) methods of studying the role of genes in human behavior is to look for patterns of behavioral inheritance in families. Traditionally, this involved applying certain behavioral assessment tests to as many members of as many generations of as many families as possible, and then applying statistical tests to the results to determine whether the transmission of these traits seemed to be heritable. While this approach has been very important in spotting potentially heritable human traits, it has suffered from questions about both the validity of the behavioral assessment tests used and the statistical methods used to define heritability. In later chapters, we see how modern molecular genetics has greatly improved the power of family lineage studies, but admittedly many questions remain.

The study of behavior in twins, and in children adopted into genetically unrelated families, has also greatly strengthened family lineage studies of the heritability of behavior. The basic strategy in twin studies, for any given variable behavior such as personality, is to look at the variability in that behavior when comparing pairs of monozygotic twins reared together or apart, as well dizygotic twin pairs reared together or apart, and to compare both of these with variability in the same trait among non-twinned biological and adopted siblings. Monozygotic twins reared apart provide additional controls for the influence of genes and environment on behavioral variability; they are essentially identical genetically, but they grow up in different cultural environments. Dizygotic twins reared together test the effect of different genetic constitutions in the same environment.

The results of such studies must also be interpreted by statistical means, and to be meaningful, many pairs in each category must be tested. When such comparisons are carried out on a large enough scale it is possible, as we will see shortly, to dissect out genetic influences from environmental influences, based on the degree to which the subjects under study share common genetic backgrounds versus common environmental influences. In working out the relative contributions of genes and environment to the differences we see between individuals, geneticists refer to a fairly straightforward formula:

$$V = G + E_s + E_n$$

This formula says simply that the variability (V) we see between two individuals should be accountable for by a combination of differences in their genes (G), plus differences in their environment (E). Environmental differences (whether cultural or ecological) can in turn be divided into those differences that are shared between two individuals (s), and those that are not shared (n). For example, two children raised in the same home would share certain environmental factors relating to parents and other family members, but would not share certain things in their external environment—different school experiences, for example, and different friends. Particularly when looking at variability involving monozygotic twins, we could also add an additional term to this equation to reflect the possible differences in the detailed aspects of fetal development discussed earlier.

We use the term "intrascale correlation," or simply "correlation," when we discuss comparisons of various individuals with respect to a given trait. Correlation in this sense is a complex parameter, dependent on sophisticated statistical arguments. For our purposes it means simply the following: a correlation of 1.0 indicates that the performance between tested pairs, carried out with large numbers of pairs, is absolutely identical; a correlation of 0 indicates the performance between tested pairs was completely random. In reality, correlations of 1.0 and 0 do not occur. Because even the same individual tested on different days rarely has a correlation of more than 0.9 on most tests, a score of 1.0 between two different individuals would not be considered real, and when the correlations of large numbers of pairs are averaged, a score of 1.0 is simply not seen. Two randomly selected individuals would by chance show some degree of correlation on most tests, although the average correlation over many pairs would seldom exceed 0.05 to 0.10. Particularly when randomly selected individuals are tested for several variables simultaneously, the overall correlation usually falls below statistical significance, but for that reason we can never say it is zero.

The Minnesota Twins Study

The case histories of the two Jimmies, and of Tommy and Bernie, are both true stories. They are more than simply anecdotal; each is one of hundreds of fully researched, well-documented stories that could be told. The details of each story vary, but enough of these studies have been carried out that we can begin to see rather clearly which aspects of human behavior are strongly influenced by genetics, and which by environment. The two Jimmies were the first set of identical twins reared apart who were enrolled in what has become the Minnesota Study of Twins Reared Apart. This program, which also studies identical twins reared together, as well as dizygotic twins reared together or apart, was founded by Dr. Thomas Bouchard at the Minneapolis campus of the University of Minnesota. Dr. Bouchard still directs the program, and has become one of the world's foremost authorities on the study of behavior in human twins.

The purpose of the Minnesota study is to document as thoroughly as possible the heritability and development over time of the physical, mental, and personality traits that make up human individuality. It is one of the largest and most comprehensive such programs in the world, but there are numerous others, both in this country and abroad. Several Scandinavian countries have twin studies that are considerably older than the Minnesota study, but with smaller numbers of subjects. In addition to having one of the largest assemblages of monozygotic and dizygotic twins (over 8,000), the Minnesota registry also has twins separated earlier, and living apart longer, than any other twin registry in the world. Included among these are over 130 pairs of monozygotic twins reared apart, more than in any other study. In the remainder of this chapter we focus largely on results from the Minnesota study, bringing in comparisons from other studies where appropriate and useful.

Identical twins reared apart are identified for the Minnesota study by a variety of means. There was considerable publicity when the two Jimmies—now known as Jim Springer and Jim Lewis—were first reunited in February 1979. The resulting media attention, laced with hints that the researchers would be interested in finding additional sets of twins reared apart, led to the rapid identification of a number of other suitable pairs of both monozygotic and dizygotic twins. Twins came to the study through self-referral, or were brought to the attention of the researchers through the mediation of physicians or other health and social welfare professionals. There are now slightly more than 130 pairs of monozygotic twins reared apart in the Minnesota study, and several sets of monozygotic triplets as well. The twin pairs are of both sexes. Although twins are still occasionally added to the registry, the conditions (largely economic) that led to the separation of twins so frequently in the 1930s and '40s have changed, and the splitting apart of twin pairs is not as frequent as it once was.

Twins entered into the study are tested over a period of about fifty hours with a battery of tests to determine their genetic relatedness and medical condition, including electroencephalograms to measure brainwave patterns. They are also evaluated extensively to determine their general psychological status. Among the instruments used for the latter purpose are four personality trait inventories, three occupational interest assessments, and four different mental abilities tests. Extensive life history interviews are used to judge the similarities and differences in the home environment in the case of twins reared apart, and to probe their attitudes toward a variety of social, religious, and philosophical issues. Twins are always tested and interviewed separately, but simultaneously, to avoid any possibility of inadvertent exchange of information.

The power of twin analyses to establish a role for genes in human behavior can be seen in results from the Minnesota twin study on personality, published in several papers beginning in 1990. There is a general agreement among psychologists that most personalities can be defined according to where individuals place along five personality "axes" (Fig. 1.1). Tommy and Bernie were at opposite ends of the "agreeableness" axis, for example; the descriptors at either end of this axis fit them almost exactly. These traits are assessed by a combination of written questionnaires, plus interviews of the subjects and of family members, all supervised by trained psychologists.

Large numbers of monozygotic and dizygotic twin pairs were analyzed for individual components that make up the five major personality axes. The results for ten of the components of the Minnesota Multiphasic Personality Inventory are shown in Figure 1.2A for monozygotic and dizygotic twins reared apart. The likelihood that identical twins shared the same personality traits was considerably greater than the likelihood that fraternal twins would share these same traits in almost every case. The correlations were then averaged for all the personality components tested (Fig. 1.2B) Monozygotic twins reared together had an average correlation of 0.46. Monozygotic twins reared apart showed a correlation of 0.45 for the same traits. What this tells us is that monozygotic twin pairs are as much alike in terms of

Trait	One pole	Opposite pole
Agreeableness:	Likeable Pleasant Friendly	Quarrelsome Aggressive Unfriendly
Conscientiousness:	Organized Responsible Dependable	Careless Impulsive Undependable
Extraversion:	Decisive Outgoing Lively	Withdrawn Retiring Reserved
Neuroticism:	Doesn't Worry Stable High Self-esteem	Worrier Unstable Low Self-esteem
Openness:	Imaginative Novelty-seeking Original	Narrow Harm-avoiding Imitative

Figure 1.1 Most human personalities can be described by estimating positions along each of the above five "axes."

personality *whether they were reared in the same or different environments.* Dizygotic (fraternal) twins reared apart showed an average correlation of 0.26, which is consistent with the fact that dizygotic twins are on average about half as alike genetically as monozygotic twins. Randomly selected individuals showed no statistically reliable correlation on personality tests.

The correlation values from monozygotic and dizygotic twin studies have been used to calculate the contribution of heredity to personality. In virtually every area of human psychological development that we would ordinarily associate with personality, including those that impact strongly on our social interactions, it has been found that, *on average*, about 50 percent of the variation among individuals is related to genetic differences. The Minnesota researchers believe this figure may be closer to 70 percent in many cases, because the personality tests used, when applied to the same individual on different days, show a correlation of only about 0.8. So if an 0.8 correlation is taken as potentially representative of complete genetic identity, then the correlations for all of the twin pairs in these tests would actually be higher than the raw scores suggest.

If 50 percent of the variability we see in personality among unrelated individuals is heritable, and thus presumably due to genes, what

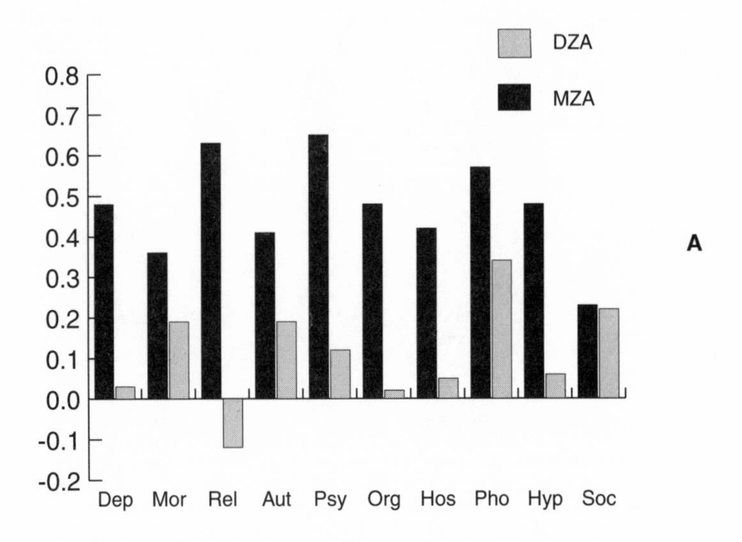

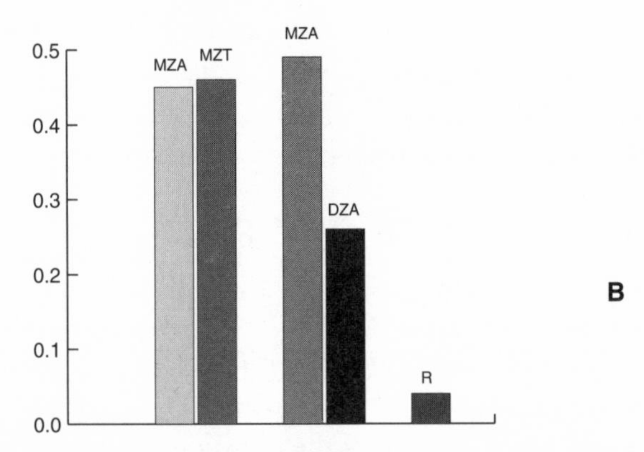

Figure 1.2 (A) Correlations for individual personality factors among identical and fraternal twins reared apart. (B) Averaged correlations for multiple personality factors for identical or fraternal twins, reared together or apart. *MZA*: identical twins reared apart; *MZT*: identical twins reared together; *DZA*: dizygotic twins reared apart; *R*: any two randomly selected individuals; *Dep*: depression; *Mor*: morale; *Rel*: religion; *Aut*: authority; *Psy*: psychoticism; *Org*: organic symptoms; *Hos*: hostility; *Pho*: phobias; *Hyp*: hypomania; *Soc*: social adjustment.

explains the remaining 50 percent or so of this variability? Is it the environment? Undoubtedly so. But remember, environmental influences can be broken down into shared and nonshared experiences. Which of these is most important in determining variability in human personality?

It is tempting to assume that the degree to which monozygotic twins reared apart differ in personality is related to the fact that they grew up in different environments, and thus that nonshared environmental factors account for these differences. This is undoubtedly true, but it is not the whole story. Remember our fraternal twin pair, Tommy and Bernie. Reared in the same ecological and cultural environment, they manipulated it in quite different ways, eliciting different responses from the same family members, for example. One could imagine that if both boys went to the same museum together on the same day, one might be very engaged with what he saw there, whereas the other was bored and extracted little from the experience. In a room full of people, one twin might plunge in and become the life of the party, while the other sought out a dim corner where he might not be noticed. Genetically different individuals are also likely to select out different sets of friends from the same environment; peer groups are an important source of environmental influence, but they must be created from the environment by individual choice.

Similarly, it is thought, identical twins must also interact with and manipulate their environments. They will select from different environmental options those things which accord most naturally with their inborn genetic predispositions, those things with which they are most comfortable, and which give them the greatest satisfaction. Twins who are verbally adept might select books, whereas twins who are aggressive and physically adept might choose sports. Both of these things would likely be present in the environments of monozygotic twins reared apart. Fearful twins reared apart may shut out many "overchallenging" experiences in their respective environments; outgoing twins reared apart might actively seek the very same experiences. Just as genetically different individuals may select different things from the same environment and treat them differently, so too may genetically identical twins reared apart select similar things from their different environments, and treat them in a similar way.

Nevertheless, the most reasonable conclusion from the studies of monozygotic twins reared apart is that nonshared environmental differences, interacting with their shared genetic constitutions, are a major contributor to the differences we observe in their personalities. We can reasonably infer that the same is true of the population generally. What then about shared environmental experiences? To what extent do they contribute to the similarities and differences we see in siblings growing up in the same environment? Surprisingly, numerous studies of twins, and both biological and adopted siblings, have shown that shared home experiences have a minimal effect in shaping the personalities of children. For a number of personality traits, there may be a small effect of shared home environment on personality in children prior to their teen years, when they tend to mimic parents, older siblings, and neighbors as part of the learning process, and as a way of interfacing with the world. But this effect almost completely disappears by the time the children pass through adolescence and leave home. The similarities in siblings reared together appears to derive mostly from their shared genetic inheritance, and not from the home environment.

The idea that parents, through their own behavior and example, play a dominant role in forming their children's personalities is one of our most cherished beliefs. It is fundamental to our notions of how cultural values are transmitted from one generation to the next, and to our belief in parental responsibility for how children turn out as adults. But cultural transmission and personality development are simply two very different things. What the data tell us quite clearly is that as children begin to interact with the larger external world, particularly in their teen years, their own genetically determined personality factors come very much to the fore. It is this largely cultural environment of the larger society, with its larger palette of choices, its greater range of stresses, that appears to be involved in testing and strengthening some elements of personality, perhaps leaving others undeveloped at key stages of personality growth. The dominant environmental element in the cultural transformation of children is for the most part their (largely nonshared) interactions with their peers, in school and in neighborhood play.*

---

*This concept and the data underlying it have been admirably summarized by Judith Rich Harris in her book, *The Nurture Assumption: Why Children Turn Out the Way They Do* (New York: Free Press, 1998).

Our current best guess is that genetic factors and nonshared environmental factors contribute about equally to the differences we see in human personality. The less alike two individuals are genetically, and the greater their differences in environmental experience, the less alike they will be in terms of personality. Conversely, while shared genetics can account for up to about half of the similarities we see in individuals, shared environmental factors contribute very little to these similarities.

Given that experience of the world is cumulative, we might expect that, to the extent that nonshared environmental experiences are a major factor in shaping human personality, over time the impact of genes on personality would decrease, and the apparent effect of the environment would increase. We might predict that the longer genetically identical twins lived apart, for example, the less like each other they would become. But this is not what we see. The degree of likeness, of concordance on personality tests, holds up remarkably well, even into the eighth and ninth decades of life. Some tests even suggest that identical twins grow more alike with time, whether reared apart or together. It is often noted that we become more "set in our ways" over our life times. Our "way," it seems, although modified by our environment, is determined to a substantial degree by our genetic inheritance.

Genes and Behavior

Farmers and other breeders of domestic animals have known for thousands of years that a wide range of animal characteristics can be selectively bred into controlled animal populations, and thus (although they certainly did not know this until the past hundred years or so) are under genetic control. Particularly with animals such as dogs, behavioral traits such as agressiveness or docility, loyalty and bravery, and even the ability to manage flocks or retrieve game, can be enhanced through selective breeding. As long as breeding is controlled, these behavioral and personality traits are passed faithfully from one generation to the next and are unquestionably under genetic control.

Results such as those shown in Figure 1.2 suggest that genes also play an important role in determining human personality. On the other hand, what does personality have to do with human behavior, which is the subject of our inquiry? And how exactly might genes be involved in determining personality? Perhaps most important, what is the underlying biological significance of human behavior? And why is it so variable?

As we see throughout the rest of this book, all behavior, in ourselves and in all other animals, consists of three basic components. First is the *perception* of something in the environment—we sense some sort of stimulus that catches our attention. Second, there is an *integration* of this perception with previous experience, for example, memory. Third, we mount a *reaction* to the stimulus. That is a very stripped-down definition of behavior, but such a definition is useful as we look at the variability in behavior we see among different organisms of the same species. Within this framework, variability in the genes governing our perceptions, and transferring this information to our brains, ultimately influences how we respond to things in the environment, whether those "things" are physically real, like a predator, or abstract, like a melody. Genes affecting the way we process the resulting signal—how our brain cells talk with each other and with the rest of the body—also affect behavior. Finally, genetic differences affecting the brain's ability to direct a response to a given stimulus could certainly alter behavior between individuals. Notice that at all three levels, we are talking about the nervous system—the brain and its collateral tissues—as the target for genetic regulation of, and variability in, behavior. Naturally occurring differences between individuals in the genes regulating any of these processes may well explain differences in the way different people react to the same external situation.

The twin studies tell us that variations in genes between individuals contribute significantly to variability in human behaviors such as personality, but they tell us nothing about the possible nature of these genes. We examine a number of genes known to be involved in various components of behavior, in ourselves and in other living organisms, in subsequent chapters. There is a long tradition in genetics of identifying individual genes based on mutations that alter an organism in some observable way. Much of the early history of human genetics was devoted to looking for single-gene differences that could explain differences in things like behavior. But research carried out over the past fifty years has made it absolutely clear that single genes by themselves are rarely—if ever—responsible for any human behavior. Even subdomains of personality, such as the components laid out in

Figure 1.1, are far too complex to be explained by single genes. Everything we understand about behavior suggests it is best explained by the interaction of many genes, with each other and with the environment, through multiple dimensions of time and space, and in ways we may not even yet be aware of.

On the other hand, single genes, when defective, can often *disrupt* human personality and behavior. A single-gene defect that results in chronic pain, for example, could cause significant changes in an individual's behavior. Huntington's disease, which is caused by a single gene, begins with major personality disturbances. The single genes which, when disabled by mutation, are known to cause the highly heritable, early-onset forms of Alzheimer's disease certainly cause major personality disruptions. However, the fact that a single defective gene can disrupt a given personality trait tells us only that that gene is involved in the trait under consideration, not that it is the *only* gene involved.

We return time and again to twin studies as we examine other areas of human behavior, because such studies offer the clearest insight we have into the role of genetics as well as environment in all aspects of the behavior of human beings. Anthropocentrism is always a danger in biology, but it is probably safe to say that behavior is more complex in human beings than in any other animal, if for no other reason than that our systems for integrating signals from the environment—our brains—are more complex. Complex systems can sometimes be more easily understood by first developing an understanding of how simpler systems work, and that is one approach we use throughout this book. Where then can we find the simplest behavior? And how do we define behavior in the simplest systems? These are important questions, because a clearer understanding of the evolution of behavior, from its earliest appearance and most primitive expression, will contribute a great deal to our understanding of behavior in ourselves. We begin this journey in the next chapter.

2

In the Beginning

The Evolutionary Origins of Behavior

When we think about behavior, naturally enough we think about how human beings react in response to the people, things, and events in their environment. Human behavior is enormously complex, involving intellectual, emotional, and social elements. Some components of human behavior, such as speech and abstract thought, are either unique to humans, or are unusually highly developed in our species.

Yet there are many elements of behavior shared by almost all living animals, including humans. Information about the environment—everything that lies outside the individual—is picked up through our five senses and processed in the brain and other elements of the nervous system, and then we respond, physically or psychologically, to those cues. We recognize similar behavioral responses in other animals—our pets, for example, or any of the wild animals that have been the subject of the seemingly endless nature programs on television. If we live closely enough with animals to become truly attuned to their daily existence, we begin to see traits in them that we identify in otherwise human terms: personality, for example, or intelligence; shyness, or guile.

The study of behavior is a science in its own right, and the branch of biology that deals with it is called ethology. It is difficult to get an ethologist to say exactly what it is that he or she studies. In terms of animal life, what is it we are defining as behavior? While there appears to be no standard definition to which everyone subscribes, we can probably use the following definition without fear of rebuke from ethologists: behavior is defined largely by whatever it is that an animal does to stay alive and reproduce. For the most part this involves avoiding predators and other forms of accidental death such as starvation; finding a mate; and producing as many offspring as possible before dying. This is a much more stripped down definition than the one given, say, in the Encyclopedia Britannica for human behavior: the sum of a person's cognitive, emotional, and social capabilities. But in fact, the endpoint of human behavior is not terribly different from the stripped-down version. We may go about it in very complex ways, but from a biological perspective our goal, too, is to reproduce our kind as often and as successfully as possible.

Ethologists study behavior in laboratory and domesticated animals, as well as the large game animals we see on television, but they also look at behavior all the way down the food chain, in organisms like insects and worms. One of the most intensely studied of these organisms, the common fruit fly *Drosophila melanogaster*, engages in incredibly complex behavioral patterns, ranging from courtship and mating to simply finding the next meal. One of the advantages of biologically simpler, rapidly reproducing organisms such as Drosophila is that it becomes much simpler to uncover the genetic patterns underlying behavior; we have a close look at Drosophila behavior in Chapter 7.

The finding of distinctive behavioral patterns in organisms like fruit flies raises an interesting question: How far down the evolutionary tree can we descend and still detect evidence of "behavior"? And what, if anything, does behavior in such organisms have to do with human behavior? In ourselves, we associate behavior with our brains and our highly sophisticated nervous systems. Indeed, ethologists also correlate virtually all animal behavior with perceptions of and responses to the environment by nervous systems. But if we follow life forms far enough back into evolutionary history, we eventually run into animals with barely recognizable nervous systems. In the earliest metazoans (multicellular animals), there may be only a single sensory cell, connected directly to cells charged with mediating whatever response is possible, without any intermediate processing of the signal. Or, as in sponges, there may be no identifiable nerve cells at all. Can we define behavior in animals that lack nervous systems?

Let's face this question head on, and step even farther back in evolution. All animal (and plant) species on the earth today are descendants of life forms referred to collectively as the *Protoctista* or "protists" (Fig. 2.1). All protists are single cells. They were the second major life form to appear in evolution, the first being the Monera. The monerans are also single cells, but they are much smaller than protists, and they are also much simpler in terms of internal structure. Monerans

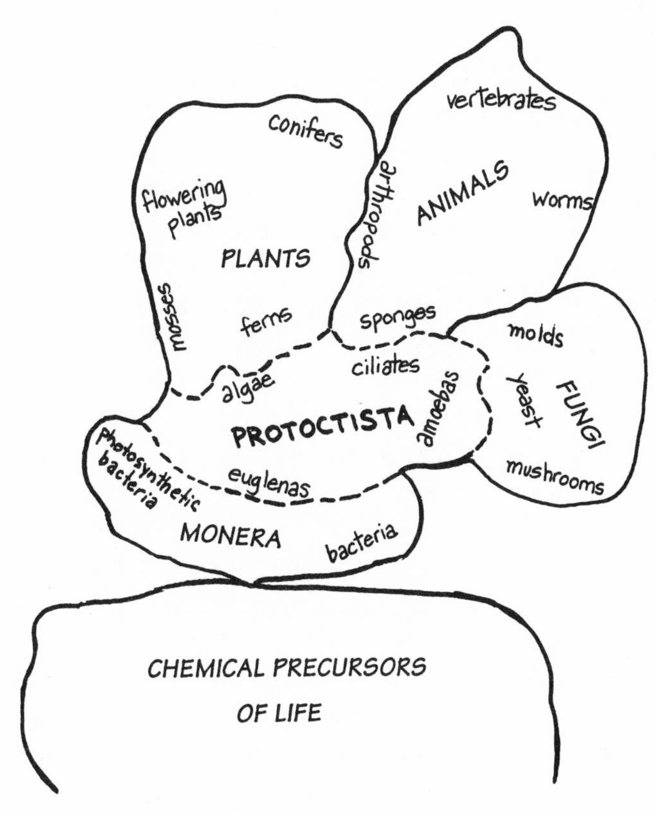

Figure 2.1 An evolutionary "tree."

are represented in today's world by the bacteria. But since protists are our nearest single-cell ancestor, we focus on them.

We ourselves are composed of single cells—somewhere between 50 and 100 *trillion* single cells, to be precise. All animals descending from the protists are composed of multiple cells, that is, are multicellular. Individual cells, whether free-living like the protists or as part of a multicellular organism like humans, are in a sense the "atoms" of life; they are the simplest thing of which we can say, "this is alive." By alive, we mean essentially something able to extract energy from the environment, and to use that energy to produce DNA-directed copies of itself, by itself. By this definition, for example, viruses (which are much smaller than cells) are not alive. Unicellular protists and the monerans are unique in that the organism *is* the cell, and the cell itself is an organism. Since all the protists are unicellular, there can be no question of involvement of a nervous system in their life histories. Can we detect anything we could reasonably call behavior in protists?

We can indeed. Let's look at just one of these creatures, an organism called *Paramecium* (Fig. 2.2), to see how it goes about its daily life, and how we might expect to encounter examples of behavior. Paramecia live in freshwater ponds, where they feed on the bacteria that in turn feed on decaying animal and vegetable matter. They are covered with several thousand tiny, hairlike structures called *cilia*, which they use for propulsion through the water. The cilia lining our nose, throat, and other internal bodily surfaces today have changed very little since they first appeared in protists like paramecia a billion or so years ago.

Protists are much larger than their moneran ancestors, and for single-cell organisms they are remarkably complex. Over evolutionary time they developed an amazing range of structural specializations that anticipate the design features of much more complex organisms. Paramecia have a readily observable structure called an oral cavity, which is a highly specialized region of the cell surface for engulfing a small sample of water (hopefully containing bacteria) and carrying it inside the cell. Simpler substances such as salts and gasses dissolved in water are exchanged directly across the entire cell membrane. Zoologists arbitrarily use the oral cavity to define the anterior end of the cell. There is also a structure toward the posterior end of the cell called the anal pore, for discharging undigested food.

Size has definite advantages. It certainly discourages predators, but

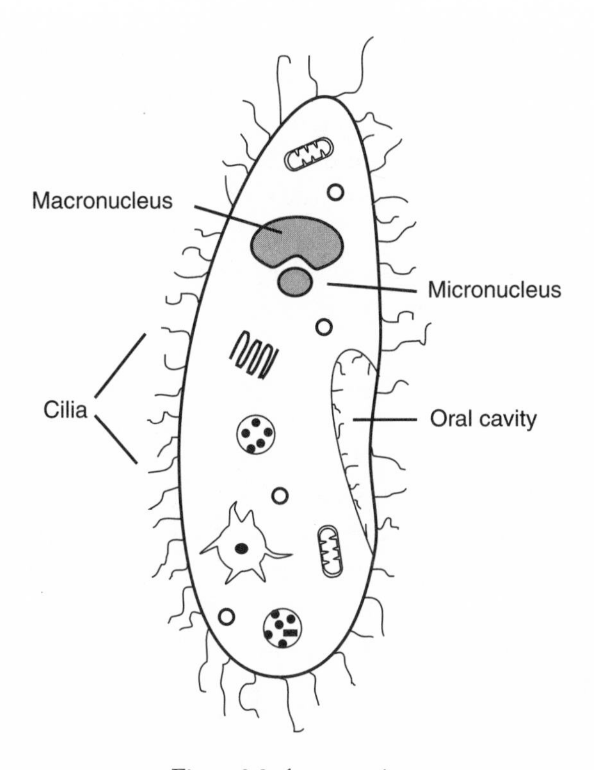

Figure 2.2 A paramecium

perhaps of more importance in the early days of life on earth was the ability to store food inside the cell for use when nutrients in the environment dwindled or disappeared. Although barely visible to the unaided eye, paramecia are easily a million times larger in volume than most bacteria, providing plenty of room for laying in supplies. But size is not without its own problems and limitations; one of these is making enough proteins to operate a cell on a daily basis. Proteins are made from information stored in DNA in the form of genes. There is a limit to how fast this information can be copied, and thus to how many proteins can be synthesized per minute or per hour. Cells can get larger, but the amount of DNA they contain is fixed. At some point there simply is not enough DNA to keep up with the cell's demand for new proteins.

To meet the demand for the extra DNA needed to operate such a large cell, paramecia created two different kinds of nuclei. The macronucleus houses the DNA used to direct the production of proteins needed by the cell to conduct its daily business of eating, respiring, moving about, and so on. The chromosomes stored in the macronucleus are replicated hundreds of times over to generate additional copies of key genes. This overcomes the limitation placed on the rate at which individual genes can be read for the production of needed proteins.

Paramecia also have a micronucleus, which contains a single, unamplified, diploid set of chromosomes. For most of the life cycle of the cell, the micronuclear chromosomes lie unread and unused. But paramecia are capable of using sex as part of their reproductive strategy. When one paramecium finds another paramecium ready to have sex, the micronuclei come into play, as we discuss shortly.

It is worth noting that in these simple, single-cell organisms we see for the first time one specialization that would become a signature feature of all higher life forms: the segregation of DNA into two functionally distinct pools. The micronuclear DNA used for reproduction in paramecia is the functional equivalent of the DNA stored in the germ cells of multicellular animals (sperm and ova in humans). The macronuclear DNA, used by paramecia to direct its interactions with the outside world (and indirectly to foster the interests of the micronuclear DNA) corresponds in effect to the DNA housed in all the rest of the cells of our bodies—the soma.

The question of whether or not organisms such as paramecia engage in anything we could reasonably call behavior was first addressed seriously around the turn of the twentieth century by a remarkable scientist named Herbert Spencer Jennings. The son of a small-town physician who encouraged his interest in the natural sciences, Jennings spent most of his life as a highly respected researcher at the Johns Hopkins University in Baltimore. It was during his formative years as a scientist that Mendel's laws governing inheritance were rediscovered, and the emerging field of genetics would became a great source of inspiration for him. Jennings began to study patterns of inheritance in paramecia and similar single-cell animals, and quickly became interested in whether or not such organisms exhibited characteristics that could be considered behavior. Darwin, among others, had raised the question of whether behavior in general was heritable. Darwin's cousin, Francis Galton, would be the first to attempt to look at this in human beings in a scientific way. Jennings saw the many advantages of approaching such questions in the smallest possible living systems, and contented himself with analyzing behavior in paramecia.*

Jennings found that paramecia are not simply little sacs of protoplasm floating in a pond. They are highly active, moving about very well through use of their cilia, which they use more or less in the way we use our arms and legs to move through water. Their motion is not entirely random. The major purpose of movement is to find food, but it can also be used to avoid danger. Like humans, paramecia are responsive to extremes of temperature, and will swim around until they find a comfort zone. They are also sensitive to touch, and like any other animal will move away when poked. They can orient themselves in a gravitational field and can respond to sound waves, characteristics that show up in the ear structures of vertebrates, for example. The ability to orient in a gravitational field keeps these cells near the top of the pond, where oxygen is more readily available. Some protists use a mechanism similar to that used by higher animals to maintain their balance in a gravitational field—a small, dense structure that sinks under the influence of gravity, touching that side of the cell closest to the earth. Whether paramecia use this sort of mechanism is not known. Paramecia can also orient themselves in a centrifugal field, presumably by the same mechanism.

Using a different mechanism, which is not understood but probably involves the sense of touch through their cilia, paramecia orient themselves facing upstream in moving water. Another orienting activity involves electrical fields. In mild electrical currents, paramecia will all line up and swim toward the cathode (the negative pole). But if the current is increased beyond a certain point, they will all turn around and swim to the anode. Although lacking photoreceptors as we think of them in connection with eyes, paramecia have photo-sensitive pigments that allow them to sense ultraviolet light. UV light is highly damaging to DNA, and organisms like paramecia, lacking any sort of protective coat, are particularly susceptible to UV damage. They are also very sensitive to chemicals in their environment, anticipating our

* Some of Galton's studies prepared the way for the emergence of eugenics, which in turn gave rise to some of the most preposterous misuses of science humanity has known. Interestingly, Jennings would lead the charge in the United States against the misuse of genetic information by Galton's successors. These issues are explored in more detail in Chapter 12 and Appendix I.

senses of smell and taste. Jennings made the point early on that, all in all, paramecia respond—however crudely—to pretty much the same environmental cues that we respond to.

Paramecia can respond to a wide range of potentially noxious stimuli by a form of avoidance behavior, which is qualitatively not different from this kind of behavior in much more complex organisms. The receptors for both chemicals and temperature appear to reside along the cilia. When they encounter water that is too hot or too cold, for example, or which contains unfamiliar chemicals, paramecia quickly reverse the direction of their ciliary beat, turn to one side, and swim off in another direction. They will repeat this action until they reach water that is acceptable to them. When the immediate environment is acceptable, they may attach to solid materials and rest for awhile, saving the energy required to operate their cilia. Cilia around the oral cavity continue to beat, however, constantly bringing water samples and food into the cell. So these organisms are able to differentially regulate the ciliary structures used for locomotion and feeding.

Protists like paramecia are among the earliest life forms to engage in sex. For most of their lives, paramecia divide asexually, by simple fission, just like their bacterial ancestors. One cell duplicates its DNA and then divides itself in half, passing a copy of the DNA and a portion of the cytoplasmic contents to each daughter cell. The daughter cells are all clonal copies of the parental cell—all members of a clone are, in effect, genetically identical twins. But unlike their bacterial ancestors, paramecia cannot repeat this process endlessly, even in the presence of an adequate food supply. After a hundred or so fission events, the clonal progeny begin to show signs of aging. They alter their shape, move around more slowly, and take increasingly longer to undergo the cell replication process. Eventually, they stop dividing altogether, and then they die—unless they have sex.

The process of sexual reproduction in paramecia is referred to as conjugation (Fig. 2.3). Individual paramecia seek out a suitable conjugal partner. Two paramecia are brought together in the region of the oral cavity, probably through the same ciliary action used to sweep food into these structures. This "preconjugal kiss" allows the two individuals to determine each other's mating type. If the two partners are of different mating types, and overall conditions are right, they then proceed to clear the cilia away from their apposing ventral regions to form

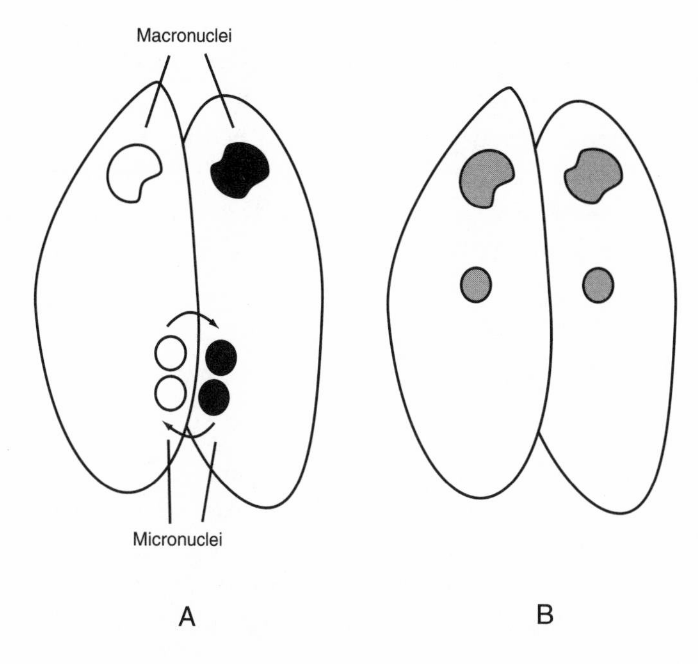

Figure 2.3 Paramecia mating. (A) Two genetically distinct paramecia (black vs. white nuclei) exchange micronuclei. (B) After conjugation, the two emerging paramecia are genetically identical with each other and genetically different (gray nuclei) from either parent.

a "conjugation patch." If they are of the same mating types, they eventually wriggle away from each other and go their separate ways.

If the two cells commit to mating, the two conjugation patches fuse tightly together. The two partners then exchange DNA—genes—across the patch, and fuse the incoming micronucleus with one of their own haploid micronuclei to create a diploid—and genetically completely new—micronucleus. The two conjugal partners then separate, and each cell begins to divide again by simple fission. But each of the cells resulting from this conjugal event, and all their clonal progeny, are genetically different from the two original conjugal partners. And each has had its replicative clock "reset" to zero; each can now begin to produce daughter cells by asexual fission for a hundred or so cell divisions before signs of old age set in.*

* Paramecia are particularly interesting from an evolutionary point of view. Not only are

So, as Jennings surmised nearly a century ago, paramecia are indeed not just little sacs of protoplasm floating in a pond. They are exquisitely attuned to their environment, and by and large can sense the same things about it that we can. If we accept the definition of animal behavior stated earlier, paramecia definitely behave. They know how to find food, and are choosy about what they eat. They are alert to danger, and can take evasive action to avoid it. They can find and identify an appropriate mate, and know how to engage in sex. And they do all of this without a brain or a nervous system.

A Word or Two about Genes

Before we examine the genetic basis of behavior in paramecia, let's talk for just a bit about the nature of genes and mutations. One of the first principles of biology at the molecular level is that all of the molecules of life—and particularly DNA—are essentially the same in all living organisms. So everything we learn about genes and mutations in paramecia is directly applicable to all higher life forms, including humans.

Genes are stretches of DNA that direct the production of individual proteins within a cell. Humans have approximately 100,000 of these genes distributed over twenty-three pairs of chromosomes stored in the nucleus of each cell. The complete collection of all the genes an animal has is referred to as that animal's genome. This same gene collection is present in all other members of the same species; they all have the same genome. But some of the genes in the communally shared genome may exist in several different forms within the species. Each individual may have only a single gene for eye color, but in the species as a whole that gene may come in several different forms—blue eyes versus brown eyes, for example. These different forms of the eye color gene arose through previous mutations in that gene that proved useful in some way to the species—or, at the very least, did not prove harmful—and so were kept. The same principle applies to numerous other

---

they among the earliest living organisms to incorporate sex into reproduction, they are also among the first to exhibit aging and compulsory death. These subjects are explored in more detail in W. Clark, *Sex and the Origins of Death* (Oxford University Press, 1996) and *A Means to an End* (Oxford University Press, 1999).

genes. Slightly different forms of the same gene floating around in a population are called alleles (Table 2.1). Not all genes have alleles; many genes within a given species are invariant among individuals. But half or more of the genes exist in multiple forms. So although all members of the same species do share a common genome, the forms of some of the individual genes making up that genome can be slightly different from individual to individual. The particular combination of genes and alleles of genes present in a given individual's genome is that individual's genotype.

Genotype is an important concept, because every biological characteristic of an individual organism is ultimately influenced by its own unique genotype. Those observable characteristics that result from the expression of particular genes, or combinations of genes within a genotype, are all part of the organism's phenotype. Some phenotypic characteristics are determined directly by the genotype—a gene, alone or in concert with other genes. But other phenotypes—and behaviors are a prime example—ultimately arise as a result of an interaction between genes and the environment. We must always keep this in mind as we look at data bearing on genes and behavior, particularly in humans where the data is of necessity indirect.

Starting with Mendel himself, geneticists have traditionally identified genes on the basis of mutations. Mutations occur when a change takes place in a particular gene that in turn alters the protein encoded by the gene. If the alteration results in a measurable change in some aspect of an individual's phenotype, *and* if this change is passed on to that individual's offspring (i.e., occurs in a germ cell rather than a

Table 2.1. *Some common terms in genetics*

Gene	A portion of DNA that contains the information for construction of a particular protein
Alleles	Slight variants of the same gene in a given species
Genome	All of the DNA possessed by an individual member of a species
Genotype	The particular collection of various possible gene alleles present in the genome of a given individual
Phenotype	The physically discernible features resulting from a particular genotype

somatic cell) then we say that a mutation has occurred. But in essence, a mutation is nothing more than the creation of a new allele of a gene. The hereditary component altered by the mutation was defined as a gene (from the Greek root for origin or descent), long before anyone knew exactly what a gene is. We now know of course that genes are written into DNA, using as an alphabet the nucleotides that are the chemical units of DNA.* But for the moment, we can think of a gene as the early geneticists did—simply as a unit of inheritance.

The Genetic Basis of Behavior in Paramecium

Can we define a role for these hereditary units—genes—in behavior? That is one of the most complex and sensitive questions in all of biology, and it is why we begin our exploration of this complex question in the simplest possible organism: *Paramecium*. In the paramecia we see behavior in its most elementary form: reflexive behavior. A stimulus is applied, and a response is elicited. There is no integration of the stimulatory signal by a brain, no comparison of the new event with information stored in a memory bank. No choices in response are possible; we do not have to worry about things like volition or reason, anxiety or guilt. The distance between genes and behavior is greatly foreshortened in organisms like paramecia. We can ask in a fairly direct way whether variations in individual genes, or groups of genes, are responsible for the variability we observe in responses to environmental cues. We can ask whether genes control behavior.

A number of behavioral mutants have been described for paramecia. Among the earliest was the *pawn* mutation, first described by Ching Kung in 1971. When *pawn* mutants encounter a noxious chemical substance, or an unfavorable change in temperature, they do not, like their wild-type counterparts, briefly back up and then move forward in a different direction until they find acceptable surroundings. Like the chess piece of the same name, they can only move forward, so they plow right ahead into a potentially dangerous situation. A mutation named *fast* behaves in the same way, except that these para-

---

* For further details on the chemical basis of genes and DNA, see W. Clark, *The New Healers: Molecular Medicine in the Twenty-first Century* (New York: Oxford University Press, 1997).

mecia continue moving forward at almost double the normal speed, and for an even longer period of time, than the *pawn* mutants. Kung subsequently found another mutant, with a phenotype opposite of *pawn* and *fast*, which he called *pantophobiac*. Normally, paramecia back away from a noxious stimulus for only about one to five seconds, before turning slightly and resuming forward motion. When *pantophobiac* mutants encounter such a stimulus in their environment, they jerk about spasmodically for a second or so, and then swim backward for greatly extended periods of time—some for as long as one minute—before resuming a forward course.

These mutations represent defects in the way paramecia move in response to an environmental signal. In looking for alterations in genes underlying this behavior, it helps a great deal to know exactly how paramecia go about moving around in the first place. What is the nature of the cellular machinery determining the direction in which paramecia move, and for how long?

In their quest for food and a mate, paramecia always swim forward; "forward" is the default state. They swim backward only in response to a potential threat in the environment. Forward motion is maintained by an intracellular "motor" driving the propeller-like cilia in a certain direction. This motor in turn is dependent on an electrical potential across the cell membrane that is regulated by energy-driven ion pumps, and by passive ion channels. The electrical potential across cell membranes, typically in the range of 100 millivolts or so, is generated by a differential distribution of electrically charged atoms ("ions") inside and outside the cell. Sodium ions (Na^+) are pumped out of the cell, and potassium ions (K^+) are pumped into the cell. The difference in concentration of these two ions inside and outside the cell accounts for the electrical charge across the cell membrane.

As long as this electrical potential is maintained across the surface membrane of paramecia, by continuing the separation of Na^+ and K^+, the intracellular motor driving the cilia will propel the cell forward. But if this electrical potential is destroyed, the ciliary beat is reversed, driving the cell backward. When paramecia make contact through ciliary sensory receptors with something potentially noxious in the environment, there is a rush of a third kind of ion—calcium ions (Ca^{++})—into the cell. The Ca^{++} binds to an intracellular protein called calmodulin (CaM), which then interacts with the ion channels (Fig.

2.4). First the Na^+ channel opens, allowing Na^+ to rush into the cell, driven only by the concentration difference. As the sodium ions begin to equilibrate across the cell membrane, the electrical potential drops, and this reverses the ciliary beat. As a result, the cell starts moving backward. The K^+ channel takes a second or two longer to open, but then it does, allowing K^+ to flow out of the cell. The efflux of K^+ restores enough of a membrane potential to start the cilia beating in a forward direction again, and the paramecium resumes its former forward motion. The Na^+ and K^+ channels then close, and ion pumps in the cell membrane pump out Ca^{++} and Na^+, while others pump K^+ back in, thus restoring the original cell membrane potential.

So that is how paramecia ordinarily deal with encounters with noxious stimuli in their environment. Contact with a stimulus triggers an influx of Ca^{++}, which in turn triggers a phased opening and closing of the Na^+ and K^+ channels. The slight lag in sequential activation of the channels results in a brief period when the cell is "depolarized," reversing the ciliary beat and allowing the cell to back away from a potentially harmful signal.

In recent years, scientists have begun uncovering the genes which, in differing allelic forms, account for the behavior of some of the paramecia mutants. The gene underlying both the *pantophobiac* and *fast* mutations turns out to be the gene for calmodulin. In paramecia, different regions of the charged calmodulin molecule interact with the Na^+ and K^+ channels. In *fast* mutants, a calmodulin mutation has occurred, selectively affecting the region of the calmodulin protein interacting with the Na^+ channel. After encountering a noxious stimulus, Ca^{++} enters the cell as usual, and binds to the calmodulin molecule. But the charged calmodulin is unable to interact properly with the sodium channel, and Na^+ never flows into the cell to begin the depolarization process. So the result of this simple mutation in a single gene is that the organism cannot back away from a potentially harmful substance—it just keeps moving forward.

In the *pantophobiac* mutation, the region of the calmodulin protein interacting with the K^+ channel is damaged, while the region interacting with the Na^+ channel is intact. So the mutated organism begins the normal response to something unpleasant by reversing its ciliary beat and backing away. But without the inflow of K^+ to initiate reestablishment of the membrane potential, it just keeps backing up.

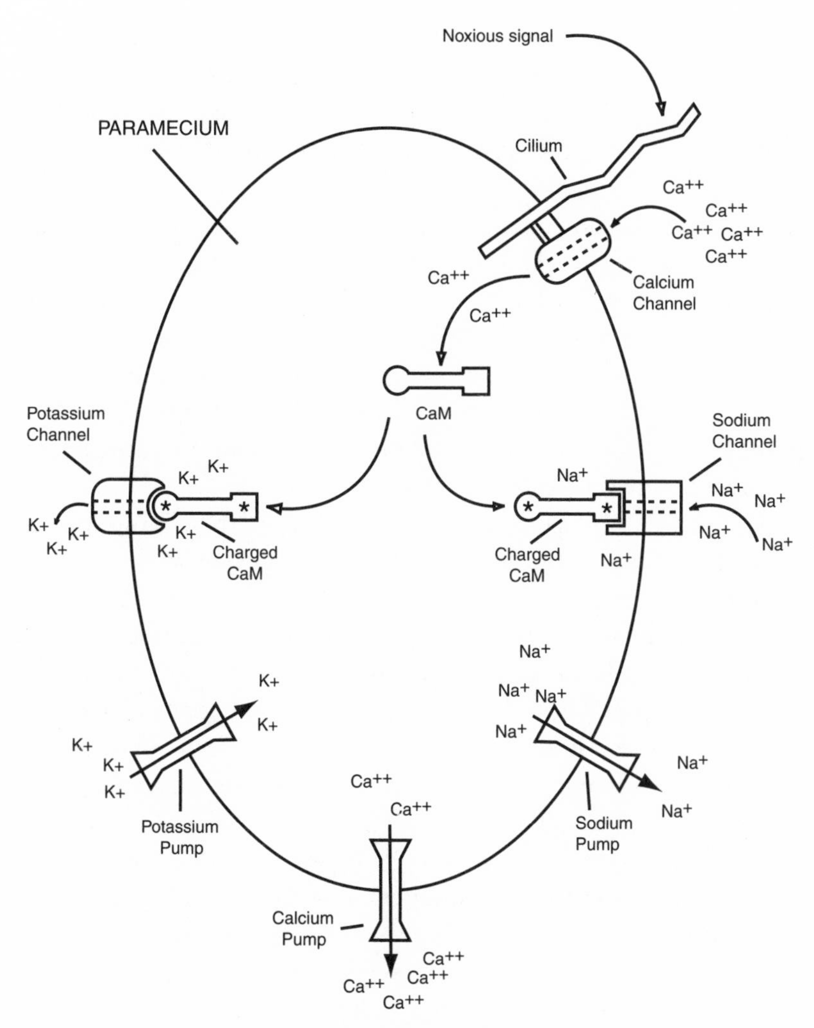

Figure 2.4 Membrane ion channels. The channels are constructed from multiple protein subunits. Depolarization occurs when the channels switch from closed to open.

Why should we care about mutations affecting the ability of single-cell organisms living in ponds to move about? What can this possibly tell us about behavior in humans? A good deal, as it turns out. The same process driving movement in paramecia—the use of Na^+ and K^+ pumps and channels to polarize and depolarize a cell membrane in response to Ca^{++} signals—is virtually identical to the process used by cells in the human brain and throughout the nervous system to generate what we call a "nerve impulse"—an electrical signal that passes from one nerve cell to another, or from a nerve cell to a muscle cell. This is the very basis of all human behavior, as we see below. The calmodulin gene found in paramecia, in only slightly modified form, plays virtually the same role in human mental processes as it does in driving a paramecium around its pond. More than one researcher has noted that the tiny, single-cell paramecia contain within them most of the salient features of human nerve cells. Later on, when life forms became multicellular, these features would segregate into separate nerve cell lineages, eventually coalescing into brain structures. In the same sense we can say of paramecia that the entire organism is both germ cell and soma, so too can we say of them that the entire organism is both nerve cell and body cell.

We should introduce here a notion that is important in interpreting much of what we have to say in this book. The discovery of a mutation in a gene that results in a change in a particular behavior, such as the calmodulin mutations affecting movement by paramecia in response to a stimulus, does not mean that the behavior in question is controlled only by that gene. What it does mean is that that particular gene is somehow involved in the behavior. The behavior itself—moving forward or backward in the case of paramecia—involves many other genes as well, for example, all of the genes that regulate the cilia that paramecia use to move about, or the genes involved in generating the energy used to drive the machinery for moving about. Very few, if any, behaviors are the product of a single gene. This is true even in single-cell organisms such as paramecia; imagine how much truer this must be in organisms as complex as human beings. That is one of the most important lessons we can take home from the study of behavior in single-cell organisms.

The fact that the behavior of paramecia is entirely reflexive does not mean it is irrelevant to human behavior. Humans also have reflexive behaviors, and they serve us quite well. A baby with absolutely no learned knowledge of its environment will withdraw its hand immediately from a noxious stimulus—a flame, for example. It may take a few milliseconds longer for a human to respond than a paramecium, because the signal must travel at least partway into the central nervous system before returning in the form of an impulse to withdraw. But such a signal need not be seen by the entire nervous system in order to produce a response; an infant does not have to make any conscious decisions about appropriate action. It is every bit as reflexive as a paramecium reversing direction in response to heat in its immediate environment. What we see in paramecia may be primitive behavior, but by any biologically meaningful definition of the word, it is behavior indeed.

So individual cells can express behavior. Do single cells in our own bodies "behave"? Some have tried to make the case for behavior of individual cells in higher animals, but for the most part meaningful behavior can be defined only at the level of the whole community of cells—at the level of the intact organism. That is where the phenotypes that we deal with in subsequent chapters are measured. And as with communities of organisms, each member of the cellular community has a defined role to play; for purposes of analyzing behavior, we find that the nerve cells play the most important role. In Chapter 4 we follow the fate of the nerve cells that retained and refined many of the mechanisms used by paramecia and other protists to sense their environment. These cells eventually coalesced into nervous systems with increasingly sophisticated sensing and reacting mechanisms. At each step we look closely at the extent to which the behavior we observe in these increasingly complex organisms can be related not only to evolving nervous structures, but to the genetic mechanisms we believe underlie all biological phenomena. We will see in every case we examine that behaviors are always the product of the coordinated actions of many different genes. But for now, let's continue our exploration of what is known about the genetics of behavior in human beings.

3

The Nose Knows

Cells communicate; they "talk" to one another. Free-living cells, such as paramecia or yeast, release a multitude of chemical signals that carry messages about food sources, possible dangers in the neighborhood, and the availability of sex. Cells of the same species have receptors that specifically recognize these chemical signals, and a cell's behavior is very much influenced by messages it receives from its nearby relatives. Free-living cells use these receptors to pick up important messages from the surrounding environment; these signals alert them to food and danger, and provide information about the general lay of the land.

When life forms became multicellular a half-billion or so years ago, cells continued to refine their ability to interpret environmental signals. They also retained and improved their ability to communicate with each other in the new context of a multicellular organism. Cells throughout the body are constantly releasing information about their own internal worlds, and eagerly gather in signals from other cells telling them how things are in the nether reaches of the organism. Chemical signals exchanged between cells in multicellular animals are called cytokines, of which hormones are one well-known subset.

As complex multicellular organisms, we humans need to know what is happening in our external as well as our internal environments. We commonly think of gathering external information through our five senses—sight, taste, smell, hearing, and touch. Actually, our sensory systems are far more complex (Table 3.1). Taste and smell are really the same kind of stimulus, involving chemicals in liquid and air, respectively, that interact with our chemoreceptors. Touch and sound are also fundamentally the same—both provoke responses of the body to physical pressure. Moreover, we also need to account for sensations such as temperature and pain, and the fact that we are sensitive to things like barometric pressure. All of this information is gathered by sensory neurons not only concentrated in specific sense organs such as eyes and ears, but scattered throughout our bodies as well.

But humans have long wondered whether the standard battery of external sensing devices described to us by scientists is really the whole story. How many of us have walked into an empty room and sensed the presence of another person, even though our "senses" tell us absolutely no one is there? How many times have we sensed danger in a situation where these same senses tell us there is no threat? Why do we meet some people for the first time and feel we know them, even though our senses pick up nothing at all familiar about them? And what is this mysterious thing between people called "chemistry"? Why do some people elicit feelings of warmth in us, while others elicit immediate and intense dislike?

Table 3.1. *Major categories of sensory receptors.*

Receptor category	Stimulus perceived
Photoreceptors	Light
Mechanoreceptors	Touch; sound; air pressure
Chemoreceptors	Taste; smell
Nociceptors	Pain
Thermoreceptors	Heat
Osmoreceptors	Osmotic pressure

Psychologists could provide us with endless reasonable explanations of all of these happenings, many of which would doubtless be true. But can we really rule out the possibility that there may be ways of communicating between human beings that have nothing to do with our standard senses? In 1971 Dr. Martha McClintock, then at Harvard University, published findings that raised some new and, some thought, rather radical questions about the way human beings gather information about their world. Her research suggested that humans may indeed possess a long-suspected "sixth sense." She found that individual human beings may be using a type of interpersonal sensory communication once thought to exist only in nonhuman animal species—pheromones.

Pheromones are defined as chemicals released by one organism into its surroundings that influence the behavior of another organism, usually of the same species, and often of only one sex. They may be released into the air or into water—either into the general surroundings, for aquatic animals, or into urine, for land animals. They are widespread among animal species, from the earliest protists and yeast up through mammals and even primates. Pheromones are picked up through highly specific protein receptors located on chemical-sensing neurons. In general, animals "tune out" the pheromones of other species (or, in the case of most pheromones related to reproduction, pheromones of the same sex in their own species) by simply not having an appropriate pheromone receptor.

Pheromones play a number of roles in animal societies, usually directing the targeted individual toward a new activity. The behaviors most often elicited by these compounds are related to reproduction or to defense. For example, pheromones released by queen bees and ants are used to regulate the activities of other members of their complex colonies. But ants also use pheromones to lay trails for other food-gathering ants to follow once they identify a food source. Rats use pheromones to warn one another about dangers in a particular food source, but also to regulate mating behavior. Cats use them to mark territory and to signal an interest in sex.

Pheromone receptors can be extraordinarily sensitive. For example, the pheromone bombykol, released by female silkworm moths, can be detected by a male moth nearly a mile away. Pheromones work at concentrations billions of times lower than is usual for chemically mediated responses such as taste or smell. As little as one molecule per second impinging on a male moth's chemoreceptor for bombykol is sufficient to cause him to change direction toward the female. Once a pheromone signal is received by a male moth, he simply stops what he was doing, orients himself in the direction of the wind, and begins to move upwind toward the female. This is basically the same behavior exhibited by paramecia moving toward a positive chemical stimulus, or orienting themselves in moving water.

McClintock stumbled on the possible use of pheromones by humans while looking at a phenomenon that had intrigued investigators for many years, but for which the evidence was at best anecdotal—that women who live closely together tend to synchronize their menstrual cycles. She studied 135 women, aged seventeen to twenty-two, living together in a single college dormitory building. The women were analyzed closely not only for the particulars of their menstrual cycles, but also for the degrees of closeness with other women in their daily environment. "Closeness" in this study was defined both in terms of physical proximity and time spent together each day, as well as in terms of emotional closeness. What she found was that these young women did indeed synchronize their cycles. Synchrony was found among women who were roommates, that is, who were physically close and spent a good deal of time together, and also among women who considered each other "best friends," that is, were emotionally close. The strongest synchronization of all was found among women who were best friends and roommates—close both physically and emotionally.

McClintock's results were confirmed in a number of subsequent studies by other researchers. Confounding factors were identified—irregular or unusually long menstrual cycles tended to obscure the effect, for example. Women on birth control pills have menstrual cycles that are determined by the hormones in the pills, and such cycles were not subject to synchronization. But most studies agreed with McClintock's basic finding, and strengthened the notion that an emotional bond seemed to be a major factor in synchrony, although clearly secondary to regular physical proximity.

There are several possible explanations for such a phenomenon. A number of things are known to affect human menstrual cycles: prolonged breast feeding; excessive loss of body fat, for example through strenuous athletic activities; clinical hormonal manipulations. None of these applied to the women in McClintock's study. On the other hand, all of these women were living in essentially the same environment, subject to similar stresses, nutrition, light and dark cycles, and so forth. These could conceivably exert a synchronizing effect on the women's biological functions. This was effectively ruled out in McClintock's study by the fact that women who were neither roommates nor best friends, but living in the same dormitory during the same period of time, showed no synchronization whatsoever.

Another possibility is that women living closely together, particularly if they interact often and closely as best friends, could be communicating through a variety of subtle social signals. Women who are close on a daily basis usually know when each is having her period. This could conceivably lead to a mutual subconscious emotional impact on one another's cycle. Numerous physiological changes can be detected when two very close friends come into contact with one another, presumably triggered by sight, hearing, and touch. An impact of this kind of communication on a bodily process such as menstruation would not seem outlandish. But there is a third possibility, one that McClintock favored from the beginning and would eventually prove: The women could be communicating by means of pheromones.

Pheromones: A Sixth Sense?

McClintock's initial results were received with great interest, but also a great deal of understandable caution. Previous suggestions that pheromones might play a role in human interactions had generally been dismissed. McClintock's experiments were better designed than some of the earlier studies, but even so most biologists scrambled madly to find another explanation. Why? What are these mysterious entities, and why were scientists skeptical about their existence in human beings?

In mammals, pheromones are secreted in urine or sweat, often in a rapidly volatilizing liquid form, and are picked up by a special structure present in nearly all vertebrates called the *vomeronasal organ*. This is a tubular structure made of cartilage, found along the lower portion of the septum dividing the two halves of the nose. Although located in the nose, and clearly related to smell (olfaction) evolutionarily, pheromone communication is actually quite distinct physiologically. Olfactory neurons interact with chemicals in the air, and conduct nerve impulses to the olfactory bulb in the brain, which lies just above the nasal cavity and connects to the olfactory cortex and various other brain regions where information about smell is processed (Fig. 3.1). There are about a thousand different chemical receptors associated with various olfactory neurons. The reactions triggered by standard olfaction may include complex thoughts and emotions as well as direct actions, but these are almost always fully conscious responses.

Vomeronasal neurons have many fewer chemoreceptors than do olfactory neurons. Moreover, olfactory neurons have cilia protruding from their ends, facing into the nasal channels. Vomeronasal neurons lack these cilia. In mammals, even airborne pheromones are sensed not in the air, but in liquid. Fluids gathered from the nose and mouth are

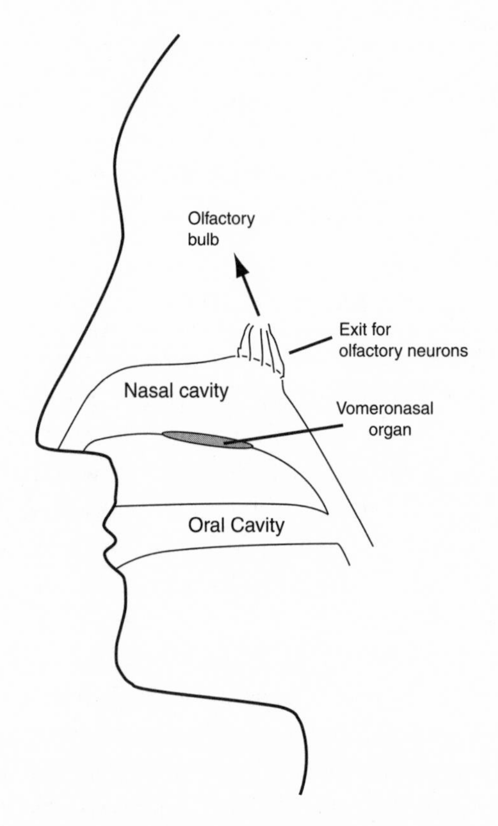

Figure 3.1 The vomeronasal organ in humans.

pumped through the vomeronasal structure, allowing pheromones dissolved in them access to chemosensing neurons. That explains why many mammalian pheromones are proteins, which are too large to volatilize but are soluble in water, and why some pheromones appear to be picked up orally, for example, when animals lick one another. Many of these proteins are found in urine, sweat, and sexual secretions. The vomeronasal neurons channel information through a separate *accessory olfactory bulb* in most mammals,* which connects to regions of the brain (the amygdala, hypothalamus, and pituitary gland) involved in controlling numerous hormones, in particular, reproductive hormones. Removal of the vomeronasal organ in mammals nearly always reduces reproductive activities. The reactions triggered via this pathway are, as far as we know, entirely subconscious.

The neurological distinction between pheromone signaling and ordinary smell is important. When we detect a particular odor through our standard olfactory system, we are very aware of that odor and integrate our awareness with previous experience related to that odor. As Proust described so very well, smells evoke memories, and indeed memory plays an important role in evaluating olfactory signals. Ordinary smells are also influenced by culture. Fifty years ago, body odors in a person who hadn't bathed in several days were considered quite normal. Today, many find such odors offensive. Moreover, smells judged "pleasant" are often determined by which perfume, shampoo, or after-shave lotion is popular at the moment.

None of this is true of pheromones. At least at the concentrations at which they are used by animals in nature, most pheromones are not detectable by chemoreceptors in the nose as a smell, and so recipients are simply not aware of their presence. Pheromones whose presence is not consciously perceived do not provoke thought, and the fact of their presence may not always be integrated with previous experience. In general, pheromones seem to trigger the expression of behaviors that have been hardwired into a given animal's system through millions of years of evolutionary conditioning.

---

* The equivalent structure has not been identified in humans; the extensive frontal regions of the cortex may have obscured it. Alternatively, the vomeronasal neurons may connect directly to the hypothalamus. It is also not clear in humans where neurons serving the vomeronasal organ exit to the brain.

In some mammals, the vomeronasal portion of the nose is very large, indicating that pheromonal signaling plays an important role in the life of that particular species. In humans this region is greatly atrophied, and was long presumed to be nonfunctional. More recent studies, however, have shown that humans do indeed have a vomeronasal organ, albeit a rather small one. It contains the types of cells expected of such an organ, and these cells can be shown to be connected neurologically to the brain. There is some evidence that these cells can be triggered by chemical signals, although much more work needs to be done. Uncertainty about the functional capabilities of the human vomeronasal structure was certainly part of the reason for the skepticism about McClintock's hunch that humans may use pheromonal signaling in relation to reproductive functions. So she and others reverted to animal studies for a number of years, trying to understand how pheromones work in mammals generally, and what they might look for in humans in particular.

Looking for Pheromones in the Laboratory

The function of pheromones in mammals has been worked out largely in connection with the reproductive habits of rats and mice. Key to rodent reproduction is something called the estrus cycle in females. Estrus in rodent females is defined by the production of eggs, by changes in the uterus, and by receptivity to males; it can occur as often as once every five to six days. The estrus cycle is like the human menstrual cycle without the shedding of the blood-rich portion of the uterine lining prepared to receive a developing embryo.

When female mice are maintained above a certain population density in the same cage, they exhibit a mutual, uniform suppression of estrus. But this effect can be prevented by removal of the vomeronasal organ. The effect of vomeronasal organ removal can itself be reversed by elevating the level of the reproductive hormone prolactin. This is entirely consistent with the fact that vomeronasal organ neurons, after stimulation by pheromones, send signals to regions of the brain that regulate production of sex hormones such as prolactin. Individual nerve cells interact with pheromones in the same way other cells in the body interact with internally produced hormones—through a specific cell-surface receptor.

Suppression of estrus in female mice induced by overcrowding can also be reversed by exposure of the suppressed females to a male or simply the urine of a male. This "male effect" can also be prevented by removal of the vomeronasal organ in the suppressed females, showing that this effect, too, is mediated by pheromones and not standard olfaction. Moreover, female mice exhibiting normal estrus can have their estrus cycles greatly accelerated by exposure to a male or his urine, and exposure to a male, to male urine, or even to certain subsets of protein derived from male urine also accelerates the onset of first estrus in virgin females. These effects are all mediated by the vomeronasal organ, and involve an alteration in reproductive hormones controlled by the brain.

Female pheromones also alter male sexual activity. Vaginal and urinary hormones from females induce mating behavior in both hamster and mouse males. In hamsters, exposure to female pheromones triggers a burst of the male hormone testosterone, which may in turn be responsible for subsequent behavioral changes in the male. If the vomeronasal organs are removed from these males, they show little or no attraction to females, and do not initiate mating activity. This is most likely related to the fact that the surge in testosterone levels following exposure to a female or to her pheromones is blocked in males lacking a vomeronasal organ.

Not all pheromones are directed at reproductive behavior. In a very interesting study, male mice were exposed to constant stress in their cage for several days, and then moved to a new cage in a different room. Untreated males, displaying the calm behavior typical of laboratory mice, were then added to the cage previously occupied by the stressed mice. The bedding in these cages was not changed, and presumably contained any pheromones secreted by the previous occupants. Within half an hour, the newly introduced mice showed elevated levels of stress hormones in their blood, and began to behave in a hyperactive, nervous fashion, just like the previous occupants of the cage that had been purposely stressed. This may be the mouse equivalent of walking into an empty room and sensing the presence of someone who is not (but may recently have been) there. At the very least it suggests that emotional states may also be communicated between animals using pheromones.

All of the effects described above, involving the action of male pheromones on female reproductive activity, can be brought about by any male, regardless of his genetic relationship to the female. But there is one effect of male pheromone that happens if, and *only* if, the male is genetically different from the female. If a female mouse in the early stages of pregnancy, prior to implantation of the embryo in the uterine wall, is exposed to the urinary pheromones of a male she is genetically unrelated to, she will abort. Disruption of the pregnancy by "foreign" male urine can be prevented if the female's vomeronasal organ is removed, or if she is injected with hormones that promote implantation. This involvement of genes in social regulation of reproductive function is one of the clearest examples we have of genetically controlled behavior through pheromones.

A basis for many of these effects was shown by Edward Boyse and his colleagues to be related to something called *histocompatibility genes*. All vertebrates have histocompatibility genes, so named because they were originally discovered as sets of genes preventing tissue transplantation between genetically distinct individuals. The mouse genes are called H-2; the equivalent genes in humans are called human leukocyte antigens (HLA). The protein products of histocompatibility genes help alert the immune system to the presence of foreign biological material in the body.

Using mice that were genetically identical except for their H-2 genes, Boyse's group tried breeding females expressing two different H-2 gene sets, with males expressing only one of the two H-2 gene types. Alert animal technicians taking care of Boyse's mouse colony reported a strange effect: the males were selectively breeding with females bearing H-2 genes different from their own. But it was not exclusive; they would mate with an H-2-identical female about a third of the time. This established a second genetically specific effect of male pheromones on female reproductive function: mating preference.

It was soon established that the male was selecting his preferred mate on the basis of an H-2-related signal present in urine excreted by the females. This was shown in a simple test. Confronted with two chambers, containing urine (only) from either an H-2-identical or H-2-disparate female, males preferentially entered the chamber containing urine from the H-2-disparate female. Males made no such distinction in the case of urine from males of these same strains, however. By using a form of associative conditioning—coupling a reward with identification of a particular urine source—it was shown that males could discriminate up to 80 percent of the time between urine from two H-2-disparate females, but they could never be trained to discriminate between the urines of two H-2-identical females. The males made exactly the same choices when the urine signal was carried only in air blown across a urine sample, showing that the signal can be airborne, as well as dissolved in the liquid urine. Because all the mice in these experiments were genetically identical except at H-2, the behavioral differences observed could be assigned unambiguously to the H-2 genetic locus.

The impact of H-2 on mate selection is not restricted to males. If the sexes are reversed in the detection tests described above, females receptive to mating will also preferentially select males whose H-2 type is different, and again, "different" means different from the H-2 type imprinted in the nest. H-2 turns out also to be the genetic factor involved in the induction of abortion in females exposed to a male genetically different from the male that impregnated her.

Histocompatibility genes are often referred to as "markers of individuality." One of their main functions is to help the body's immune system to detect bacterial and viral infections. They also trigger rejection reactions when tissue from one animal is transplanted into another. Many biologists believe they originally evolved to help members of the same species recognize and avoid mating with other members whose genetic makeup is too like their own. Such inbreeding often results in the accumulation of defective genes within the same individual, which can be fatal. How any of these functions relates to pheromonal activity of histocompatibility proteins is unclear. Whether fragments of the histocompatibility proteins themselves act as pheromones, or whether pheromonal substances that somehow associate specifically with these proteins are responsible for hormonal effects, is also uncertain. But an extremely close correlation of histocompatibility products with pheromonal activity is beyond dispute.

Several recent studies suggest that human histocompatibility (HLA) genes may also play a role in mate selection. One study looked at HLA genes in a survey of married couples in a closed religious community in South Dakota. Because all of the members of this community are descended from a small number of founding members, with little or no immigration from outside, the number of different alleles of HLA genes circulating within the community is relatively small. By typing blood samples over several generations, it was possible to establish what the HLA types of married couples in each generation ought to be if they selected among one another randomly for mates. Surprisingly, the data show a substantial bias favoring *avoidance* of mates of the same HLA type. That is, young people somehow tended to select mates whose HLA type was different from their own, even though no information about HLA types was ever made available to them prior to (or even after) marriage. Although indirect, these data strongly suggest the presence of a biological cue indicating HLA type, which influenced courtship behavior.

A more direct indication that humans respond to some sort of olfactory cue controlled by HLA type is provided by a recent experiment carried out in a group of 121 men and women who were asked to score the odor of T-shirts worn by other subjects on a scale of pleasant to unpleasant. The results showed a clear dislike for T-shirt odors from subjects with similar HLA types, and a milder dislike of or neutrality toward odors from T-shirts worn by individuals with strongly different HLA types. Moreover, respondents most often associated the less repulsive T-shirt odors with their own mates, suggesting, as did the study just cited, that HLA may have been a factor in mate selection.

Pheromones in Humans and Other Primates

To gain a clearer insight into what might be happening in menstrual synchrony in women, McClintock herself carried out some interim experiments in rats. When female rats are housed together in the same cage at normal densities, they—like human females—synchronize their estrus cycles. In a cleverly designed experiment, McClintock stationed cages containing female rats in a linked wind tunnel. In the upwind cage, she placed rats that were confirmed by examination of vaginal smears to be either in the pre-estrus or post-estrus stages of their cycles. A constant stream of air was passed through this cage into the downwind cage containing "target" rats.

What McClintock found was that air from a cage containing pre-estrus rats caused a significant lengthening of the cycles of the downwind target rats. Air emanating from a cage containing post-estrus rats shortened the cycles of the target rats. As a result of these studies,

McClintock proposed that the synchronization of estrus cycles in rats occurs not through a single pheromone, but through the interaction of two distinct pheromones, each with opposite effects on the length of the cycle. Whether either or both of these signals are related to rat histocompatibility genes has not yet been determined.

Pheromones are also used for social communication among most if not all of the primates. As with other animals, one of the major uses of pheromones is in the regulation of sexual behavior, for example, suppression of ovulation between females. Pheromones can play both indirect and direct roles in suppression of ovulation. In baboons, for example, a dominant female inhibits ovulation in other adult females in her troupe through physical harassment. She can be seen chasing them around, biting them and pushing them, taking their food away, and generally making their life miserable. This induces a state of psychological stress in the other females.

This aggressive activity toward subordinate female baboons is at its peak in the period immediately preceding the subordinate female's ovulatory period, and detection of this state by dominant females is through pheromones. These pheromones are released by all females, presumably as an attractant to males. In this case the signal is intercepted by the dominant female (who establishes her dominance in the first place by similar aggressive behavior). However, the dominant female does not suppress ovulation in the other females by releasing a suppressive pheromone. As a result of constant harassment, stress hormones of the glucocorticoid type are induced in the subordinate females, and this is sufficient to suppress ovulation. As with primates generally (including humans), stress reduces reproductive activity; in the harassed baboon females, this takes the form of suppressed ovulation. This is an example of an indirect action of pheromones; they are not directly involved in this suppression, but elicit a behavior in the dominant female that results in suppression of ovulation in other females with whom she is in contact.

But in many groups of primates, for example, marmoset monkeys, ovulatory suppression occurs without the display of aggressive or harassing activity. These primate groups are usually characterized by what is termed a "cooperative" breeding pattern, in which only one dominant female in a social group (three to five adults of each sex, plus offspring) will produce young. The other (subordinate) females assist the dominant female in rearing her young. The dominance of the breeding female is maintained without harassment or other abuse of her subordinates, yet may be even more effective than in the more socially aggressive baboons. Suppression of ovulation is not mediated by induction of stress hormones in the subordinate female, but is dependent on pheromones released by the dominant female. Here, too, initial dominance is established by competition that can be very physical and aggressive. However, once dominance is established, the resulting "cooperative" behavior of the other females is maintained without further violence. The suppressive pheromone released by the dominant female works in a direct fashion, by altering the reproductive hormones in her female subordinates.

That pheromones may also regulate reproductive function in humans was suggested in another study by Martha McClintock. In a paper published early in 1998, McClintock tested her two-pheromone hypothesis, derived from her studies on rats, directly on women. She chose for her study a group of twenty-nine women aged twenty to thirty-five, who were not on birth control pills, to act as "recipients" in her trial. She established a second group of nine women to act as donors. The donors were directed to bathe or shower daily without the use of perfumed soaps or toiletries, and then to wear cotton gauze pads under their arms for eight hours. Underarm secretions were used in these studies because all of the known glands involved in pheromone production in animals can be found at this site in humans. Moreover, in humans axillary secretions are odorless until puberty, suggesting major chemical changes related to reproduction.

At the end of each day, the pads were collected and cut into smaller squares. The smaller squares were then rubbed lightly just under the nose of the recipient women each morning, and they were asked not to wash this area for the next six hours. This procedure was carried out daily over a two-month period. The results were virtually identical to those obtained in rats. Materials collected from the underarms of donors just before ovulation ("pre-estrus") shortened the time to the next ovulation, and shortened the length of the menstrual cycle, in all recipients. Materials collected from donors immediately after they had ovulated delayed ovulation in the recipients, and lengthened their overall cycles.

These astonishing results provide the strongest suggestion to date that humans can communicate through pheromones. In these studies, odorless compounds taken in through the nose were used to communicate information about the internal physiological state of one human being to another, and to alter the recipient's own internal reproductive physiology. It follows precisely a pattern established for a pheromonal system of reproductive regulation found in rats. To establish definitively that the effect observed in humans is truly pheromonal, we would have to test it directly in women in whom vomeronasal function has been disrupted. Alternatively, we could gain indirect information by applying the same test in women with impaired olfactory function.

McClintock's most recent findings could provide an explanation for the phenomenon of menstrual synchrony she first described nearly thirty years ago, and it may well be the basis for some extraordinary results published recently by Dr. Winifred Cutler. Dr. Cutler had previously shown that women whose menstrual cycles were unusually long, or highly irregular in length, achieved menstrual regularity and normal cycle length after several months of thrice-weekly exposure to male underarm secretions. This is highly reminiscent of the studies described earlier, in which female mouse estrus cycles were significantly perturbed by exposure to male pheromones.

In her experiments, Cutler found that, as reported by their male partners in a double-blind test, women exposed to pheromones extracted from these secretions (and subsequently synthesized in the laboratory) increased their sexual receptivity and activity. The men were given a coded vial containing either highly concentrated male pheromone or an inert substance, which they then added to their aftershave lotion or cologne. A detailed comparison of their sexual "successes" with women before and after the trial showed that men using the pheromone experienced much more sexual activity with their women partners than those issued the inert substance. While it was not possible in this study to exclude rigorously the possibility that the men were themselves stimulated to greater sexual efforts by the pheromone, there was no self-reported increase in the rate of masturbation by men receiving the concentrated pheromone preparation. In any case, it is clear that inter-partner sexual activity was directly increased by male-derived pheromones.

While the findings of McClintock and Cutler must await further verification before they can be considered definitive, they will likely stimulate a veritable frenzy of research in the laboratories of pharmaceutical companies, and the identity of the chemical(s) responsible for these effects should soon be known. Will they be pheromones, as that term has been used in a strict biological sense? We will have to wait and see. They certainly have many of the characteristics of pheromones; they seem to be acting at a subconscious level, modulating fundamental reproductive behavior. And it now seems that humans cannot be dismissed as lacking a vomeronasal organ, although we don't know that the chemicals we are talking about actually work through that organ.

Whatever their nature, if these chemicals can be isolated, their potential application in helping to manage any number of ovulatory and reproductive problems in women could be enormous. Beyond immediately practical applications, however, their identification could also have important implications for the genetic basis of behavior. In those cases where genes involved in pheromone production are found, we will doubtless also find that these genes have different alleles within the population. The products of these alleles will likely result in differing abilities to produce and secrete pheromones, and in turn partially determine the effectiveness with which individuals carrying these various alleles can communicate information to others of their species. There will also be genes encoding receptors for pheromones, and these genes, too, are likely to be present in allelic form. Depending on the efficiency of the various allelic forms these receptors, individuals will be either more or less capable of receiving potentially important information about their social surroundings. These genetic differences could very well have a major impact on individual behavior.

Unquestionably, pheromones play a much more important role in the lives of animals than they do in humans. Eyesight and language play more important roles in human interactions, replacing to a great extent the need for chemical communication. We do not yet know to what extent humans normally communicate through pheromones in their daily lives, or whether there may be other pheromones, also affecting reproductive behavior or perhaps other aspects of human physiology, in the human pheromonal repertoire. There is evidence to suggest that human infants, like other mammalian offspring, are guided in their nursing behavior by pheromones released through the nipple. The possibility that pheromones could play an important role in human communication must now be taken seriously. In other mammals, many different pheromones are involved in regulating behavioral traits such as dominance relationships between individuals of the same species, recognition of other members of the immediate family or social group, or posting information about individual territories. Could the vomeronasal organ be the seat of our own sixth sense? Are pheromones the basis for that certain "chemistry" between two people? We can be sure, after the very recent results obtained by McClintock and others, that the whole question of possible chemical bases of human communication and their impact on human behavior will receive renewed attention.

4

As the Worm Turns

Learning and Memory in the Roundworm C. elegans

It took a long time to shed the unicellular way of life; the unicellular protists dominated the world for at least two billion years before giving rise to other life forms. Some protists, like the paramecia, had grown to be about as large as single cells can, and they had become enormously complex. We can see in them the beginnings of what would, in multicellular animals, become separate reproductive systems and nervous systems. They had adapted to oxygen, and had become efficient predators of their prokaryotic forebears and, in some cases, of each other. But the exploitation of size as a competitive advantage had come to something of a dead end, because of the limitations on just how large a single cell can become. One imagines that the two billion years of the protist reign were spent fine-tuning the single cell design, adding more and more internal features and external specializations. But the way to the future—to dominance not only of the seas in which life arose but of all the other biological niches of the planet—would depend on one critical event: the emergence of multicellularity.

The precise events leading up to the appearance of multicellular life forms can only be guessed at. Had paramecia simply established an internal membrane between their macronucleus and micronucleus, they could have formed a bicellular precursor of multicellular animals, with the germline/soma distinction neatly in place. But we have no evidence that this ever really happened. It seems more likely that multicellularity evolved through the coming together of individual unicellular organisms to form multicellular aggregates or colonies. The cells of sponges, which are among the most primitive multicellular animals still alive today, look very much like cells in transition from a solitary to a colonial lifestyle.

Unicellular organisms evolved through competition with other single cells for survival and for the right to reproduce. For cells exploring the possible advantages of multicellularity, this tendency to self-centered competition had to be overcome. Once the decision to become multicellular was made, some individuals had to give up their right to reproduce in order to become somatic cells, and support the reproductive rights of others—those lucky enough to become the germ cells. From the point of view of natural selection, which acts only on the basis of enhancing the reproductive efficiency of individuals, it is not immediately obvious how this could be achieved. If multicellularity were to succeed, it would also require the development of sophisticated strategies for communication among individual cells forming a colony, in order to avoid uncoordinated, chaotic interactions that could doom both the larger structure and the individuals contributing to it. The new, multicellular organisms would have to learn to behave in response to internal signals, as well as in response to the environment.

One of the great advantages of multicellularity, in addition to making it possible to become larger, is the ability to segregate different biological functions into separate cell lineages. When multicellular life-forms finally emerged from the protists about 800 million years ago to form the kingdom *Animalia*, not only did reproductive and somatic DNA segregate into separate cell lineages; the nerve-like features of single-cell protists such as paramecia also split off into distinct cell lines. It would no longer be necessary to involve the entire organism in the perception of, and response to, external stimuli. That would become the task of highly specialized cells within the new nervous systems. Behavior, which is the subject of our inquiry, is entirely a product of the nervous systems of multicellular animals.

One of the most intensely studied of the early multicellular animals still surviving today is the nematode roundworm *Caenorhabditis elegans* (Fig. 4.1). Roundworms are found in the soil and in sea sediment, but also inside people: trichinosis and elephantiasis are human diseases caused by nematodes. Roundworms (not closely related to earthworms) are one of the more primitive forms of animal life still on earth today—they were the first life form to have a complete digestive system, for example—but they have been extraordinarily successful; they may, in the aggregate, be the most plentiful form of multicellular animal life on the planet.

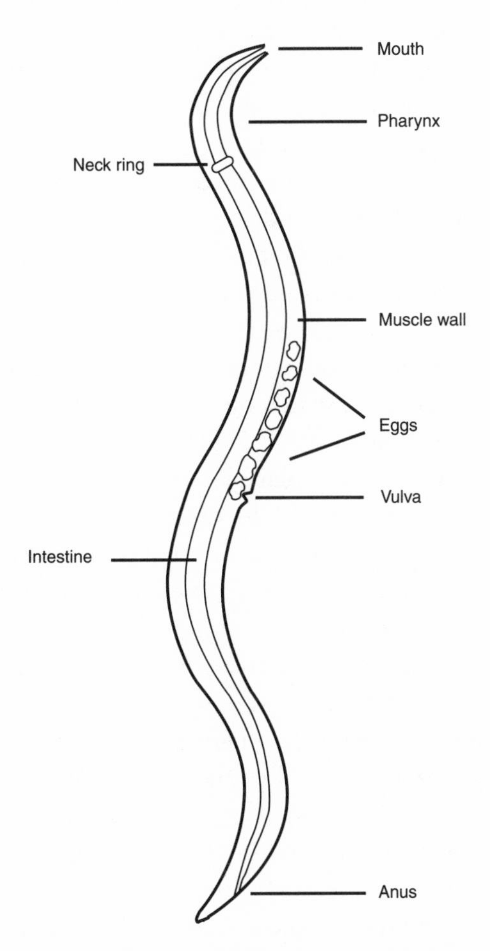

Figure 4.1 The roundworm *Caenorhabditis elegans.*

The tiny *C. elegans* worms display many of the same behavioral features as do paramecia, but in *C. elegans* we begin to see the complexities in behavioral responses to environmental cues that are typical of animals that have a nervous system. Yet the underlying cellular mechanics of these responses, and their genetic and molecular control, differ very little between *C. elegans* and paramecia. In fact they differ only modestly between either of these tiny creatures and human beings.

The value of *C. elegans* as a model system for many different aspects of eukaryotic biology was first realized by the British scientist Sydney Brenner. Brenner is a South African who migrated to England, where he received training in the emerging field of molecular biology just as it reached the crowning achievement of its early years: the unraveling of the structure and function of DNA, and the deciphering of the genetic code. Working at Cambridge University, Brenner had himself contributed to our understanding of how triplets of the nucleotide building blocks of DNA are used as letters of the genetic alphabet to store information. Others would continue to sketch out in ever finer detail exactly how genes work at a mechanistic level, but by the late 1960s Brenner began looking beyond the chemistry of genes; he was interested in the biology of genes, and in particular in how genes could guide the development and operation of a complex multicellular organism. So he set about searching for the simplest complex organism he could successfully rear in a laboratory setting.

In the early years of molecular biology, scientists interested in the genetic control of life processes had worked almost exclusively in bacteria, viruses, and yeast. Multicellular organisms were considered too complex, too difficult to grow, for routine molecular work. Unicellular organisms are ideal for genetic studies—they are simple to grow, reproduce rapidly, and have relatively few genes. They are still used rather intensively to this day for the study of a number of fundamental biological processes shared by all living cells. But by the mid-1960s, there was a growing sense that the experience and technologies gained in working out the molecular biology and genetics of microorganisms might now be profitably turned to more complex systems. Brenner wanted to move beyond the lives of individual cells, and begin a study of the genetic regulation of more complicated multicellular and intercellular processes. He wanted to know, among other things, how genes could specify the structure and function of something as intricate as a nervous system.

Genetics requires the transmission and segregation of traits over many consecutive generations, so Brenner was looking for an organism with both a nervous system and a short generation time. *C. elegans* seemed ideal. It is very small (about one millimeter in length), and reaches sexual maturity in just over three days, with an average life span of about thirteen days. Its natural habitat is soil, but it is easy to maintain in the laboratory where it feeds, like paramecia, on common bacteria such as *Escherichia coli*. Brenner published his first detailed genetic analysis of *C. elegans* in 1974; today, we know the entire sequence of the *C. elegans* genome. Worms have six pairs of chromosomes, compared with twenty-three pairs in humans. These chromosomes are rather small, containing about as much DNA *in toto* as one average human chromosomal pair. Yet worms have almost 20,000 genes, compared with an estimated 70,000 or so in humans. This is clearly a very compact genome. Remarkably, most of the genes identified in *C. elegans* have readily recognizable counterparts in humans, and many human genes, transplanted into *C. elegans*, seem to drive the same functions.

The value of *C. elegans* as a system for studying multicellular processes has been recognized by researchers in a number of different fields. Most important, developmental biologists began a study of the generation of adult organisms from the initial egg, and have now traced the origin and fate of every single cell throughout the entire life history of *C. elegans*. This is the only multicellular organism for which such extensive cellular "fate mapping" has been carried out. We now know where every single nerve cell is located in *C. elegans*, what it is connected to, and in most cases, what it does. An adult *C. elegans* worm contains a total of 959 cells, of which 302 cells are separate and distinct neurons. These neurons fall into three broad classes: afferent or sensory neurons, that perceive environmental or internal cues; interneurons, that handle communications between two or more neurons; and motor neurons, that interact with other cells in the body, triggering an appropriate response to perceived environmental or internal cues. Motor neurons activate either muscle cells, stimulating bodily movement or contraction of internal organs, or glands, triggering the release of hormones or other chemical mediators. This tripartite organization of the nervous system is typical of all higher animals, including humans. There are a total of about 5000 connections between and among the 302 nerve cells of *C. elegans*, and nearly all of these have been identified as well.

The advantage, foreseen by Brenner, is that with only 302 neurons in its entire nervous system, it might well be possible to correlate genetic mutations affecting neural function with specific neurons, allowing unprecedented insight into how a nervous system actually works—among other things, how it generates learning, memory, and behavior. Learning is the process by which information about an organism's immediate environment is imprinted on the nervous system; it is how the organism acquires information about both its external and internal worlds. Memory is the process by which that information is stored and retrieved for future use or reference. As we see in Chapter 7, memory involves actual physical and chemical changes to individual neurons as a result of their engagement in transmitting information about the environment. Behavior is the result of coordinated firing of motor neurons, which guides the organism's movement and general function within the three-dimensional space of its world. Thus at the cellular level, behavior can be thought of as the end product of information processing by the nervous system. And the nervous system, through individual outward-facing sensory neurons, provides the means by which the environment can alter an organism's behavior. So to the extent that genes are involved in affecting behavior, it will be those genes that affect the functioning of the nervous system that are most likely to be involved.

As with all of the simpler multicellular animals, the vast majority of sensory neurons in *C. elegans* are at the surface, making direct physical contact with its environment. Interestingly, some of these neurons, particularly those involved in detecting chemical substances in the environment, have extensive cilia at the end facing the outside world. These cilia are essentially the same as the cilia used by paramecia to move about. In *C. elegans* and other multicellular animals, ciliation of nerve tips provides more surface area for the expression of receptors that pick up external signals. Moreover, all three classes of neurons in *C. elegans* are driven by exactly the same kind of ion channels and pumps found in paramecia.

Many of the reflexive behaviors of paramecia are carried over into

C. elegans. For example, some of the sensory neurons of *C. elegans* are sensitive to touch. Like paramecia, if poked at one end, the worms will move away from the point of contact. Movement does not occur through the use of cilia, however. Physical contact generates the same sort of opening and closing of ion channels that we saw in paramecia, but no motor reaction takes place in the sensory neuron per se. Rather, the ionic changes are transformed into a nerve impulse, which is transferred through varying numbers of interneurons to a motor neuron, which connects to a muscle cell. It is within the muscle cell that the organismal reaction to the stimulus finally takes place. Through a reaction not dissimilar to the one driving a ciliary pump in paramecia, the muscle cell stimulated by the motor neuron is made to contract. In fact, several groups of muscle cells contract sequentially, driving the worm forward or backward with serpentine, wave-like movements.

The reaction of *C. elegans* to touch is a form of purely reflexive behavior. But the worms can go beyond simple reflex. If they are touched at one end, they always move away. If touched again on the same end, they will move away again. But if we keep doing this, repeatedly over a short period of time, the worm ceases to respond by moving away. Now, worms clearly cannot think, at least not in the way we think, but it is almost as if the worm, having been touched repeatedly without any lethal consequences, had "decided" that touching did not pose a threat, and stopped wasting energy on costly forms of avoidance behavior. This is referred to by neurobiologists as a form of *habituation*. Habituation has never been reliably detected in single-cell protists. It represents a new level of behavior, one seen for the first time in multicellular animals, and is a form of what is called nonassociative learning. It is specific, in the sense that the worm's response to other forms of stimuli—the ability to recoil from a mild electric shock, for example—is not in the least compromised by habituation to touch.

Where in the worm does this learning take place? Based on what we know of brain structures in higher organisms, we might look to the *nerve ring*, a structure in the neck region of *C. elegans* (Fig.4.1). About three-quarters of all the nerve cells in *C. elegans* send at least one nerve fiber to the nerve ring, where connections are made with one or more other nerve fibers. This begins to resemble structures like the ganglia (small nodes consisting of nerve cells) or the primitive brains of evolutionarily later organisms, where the very important process of *signal integration* takes place—the comparison of information among and between neurons bringing information from different parts of the body.

Nerve cells are, like the unicellular protists, electrically charged, or polarized; that is, they have pronounced ionic gradients across their surface membranes. Calcium and sodium ion concentrations are very low inside nerve cells, whereas potassium ions are at a high concentration, compared with the concentrations of these ions outside the cell. This creates an electrical potential which can be used to propagate nerve impulses to other nerve cells or to muscle cells. Neurons connect with each other through slender extensions—the nerve fibers—that are of two types. *Dendrites* carry information to nerve cells, and *axons* carry information away from them (Fig. 4.2). Most neurons have only one or at most a very few axons, but often have many dendrites. Moreover, both axons and dendrites can have branches, and each branch can make contact with a separate neuron. Some nerve cells in the human brain can have 100,000 or more dendritic tips. So it is possible—indeed usual—for one neuron to be in direct communication with a great many other neurons. In *C. elegans* this takes place only in the nerve ring.

Axonal branches from one neuron may connect to the dendrite of another neuron, or to the surface of a muscle cell. Neurons communicate with other cells by sending a nerve impulse across a specialized region called a *synapse*. The nerve impulse in most cases is transmitted chemically. For example, when a sensory neuron is stimulated by being

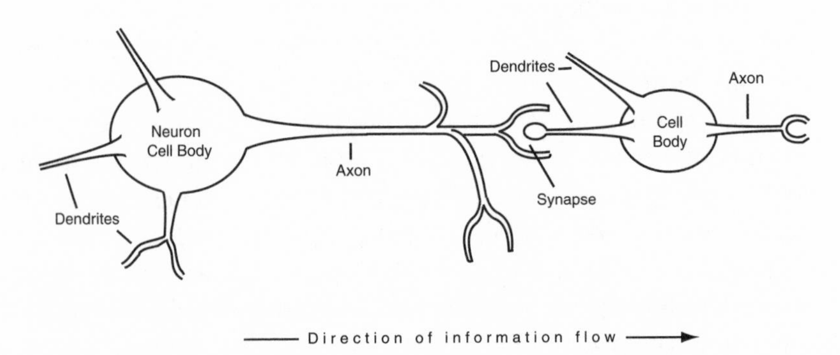

Figure 4.2 Two neurons forming an axon–dendrite synapse.

touched, it is immediately depolarized by allowing the ionic concentrations across its surface membrane to equalize. These changes are mediated by the same type of ion channels we saw in paramecia. Depolarization results in release of a special chemical, called a neurotransmitter, from the tip of the axon. Nerve cells use many different neurotransmitters to communicate with one another, and below we discuss the various ones used as we encounter them. One of the major neurotransmitters between nerve cells and muscle cells in *C. elegans* is a tiny molecule called acetylcholine. The worm form of this neurotransmitter is identical with the acetylcholine used in human nerve cell communication, even though *C. elegans* and humans are separated by nearly a billion years of evolutionary history. This is but one of many examples of the tremendous conservation of the structure and function of animal nervous systems, the principles of which were established very early in evolution.

The neurotransmitter diffuses across the synaptic space, and is picked up by receptors on the other cell. The neurotransmitter then triggers a reaction in the receiving cell. If the receiving cell is another neuron, that neuron also depolarizes and in turn releases more neurotransmitter from its own axon, thus passing the nerve impulse on to another cell. If the receiving cell is a muscle cell, then the neurotransmitter induces the muscle cell to contract, and carry out various forms of work.

One of the most important advantages of *C. elegans* is that since every single neuron has been identified, it has been possible to destroy each of the neurons one by one in order to observe the resulting defect. This is done with a combination of Nomarski optics, which allows a standard light microscope to be used to identify individual cells in a living worm, and lasers. An extremely fine beam of laser light is focused through the microscope and onto the nucleus of a target neuron. With practice, a few short bursts of intense laser light can be used to destroy individual neurons. By observing the resulting neurological deficit in worms missing a single neuron, it has been possible to determine the behaviors to which each of these cells contributes.

When many neurons are brought together in complex structures such as brains, or in the *ganglia* that run along our spinal cords, and presumably even in simple structures like the nerve ring of *C. elegans*, they have an opportunity to influence one another. Stimulation of one neuron can affect any of the other neurons to which it is connected, increasing or decreasing their ability to be depolarized by signals arriving from other sources. Stimulation of a neuron may also induce it to send out projections and establish connections to neurons with which it had previously not been in communication. Since most interneurons have dendritic inputs from many other nerve cells, the precise response of any one interneuron to a given signal is subject to numerous simultaneously delivered modulating signals. We imagine that this is the basis of signal integration, and that it is the determining factor in something like habituation or nonassociative learning.

In addition to touch, *C. elegans* worms are also responsive to chemical stimulation and heat. Temperature and chemical receptors are located at various points around the animal's surface, and represent the outward facing portions of particular sensory neurons. Eleven pairs of neurons allow worms to distinguish among hundreds of different chemical compounds. The worms are attracted to some chemicals, including certain salts, vitamins, and amino acids, and repelled by many other substances, including too high concentrations of some of the very things that attract them. Other chemicals act as pheromones, inducing mating or egg-laying behavior. As with taste and smell in higher organisms, it is often the case that several chemical receptors, with slightly different specificities, may react to the same stimulus, albeit with differing strengths.

The Genetic Basis of Behavior in C. elegans

A number of mutations have been detected in genes controlling behavior in *C. elegans*, and the characterization of these genes has advanced our understanding of the genetics of behavior well beyond what had been learned from the paramecia. Mutations affecting the ability of any of the sensory neurons to function properly could certainly impair the ability of the worms to sense their environment; mutations compromising motor neurons also affect the ability of the worms to respond to events in their environment. But understanding how genes function even at this very basic level can tell us a good deal about how the nervous system functions. For example, we know a great deal about the genetic basis of chemoreception in *C. elegans*, which is analogous to taste and smell in humans. The eleven pairs of chemosensory neurons in worms contain hundreds of different chemical receptors. Some chemicals attract worms, and some chemicals repel them.

How could attraction and repulsion be managed at the neuronal level? It could be that some chemosensing neurons, containing only receptors for chemoattractants, are hardwired into motor circuits that cause the worm to advance toward a stimulus. Other neurons, containing only receptors for repellents, might be directly connected to the machinery causing a withdrawal response. Alternatively, all chemosensory neurons could have both types of receptors; depending on which type of receptor is occupied, the same neuron could elicit a different type of motor response. Answering this question became possible with the isolation and cloning of the chemoreceptor genes of *C. elegans*. One such gene, *odr-10*, controls the perception of a chemoattractant chemical called diacetyl. Researchers found that by controlling the expression of *odr-10* in different neurons, worms would either advance toward or withdraw from a diacetyl source. So *C. elegans* has apparently concentrated receptors for chemicals it dislikes into one set of neurons, and receptors for preferred chemicals into a different set of neurons. Once stimulated by any chemical able to interact with one of its receptors—even an artificially introduced receptor—the neuron triggers only a single kind of response. Whereas paramecia were forced to develop an attraction or repulsion response within the context of a single cell, the multicellular *C. elegans* has the luxury of placing these responses under the control of separate sensing systems.

The study of genes and mutations has been extended to other forms of behavior as well. A recently described gene controls an interesting aspect of social behavior in *C. elegans*. When various strains of *C. elegans* kept in laboratories around the world are examined for feeding behavior, two phenotypes are observed. Some worms, when plated onto a Petri dish containing bacteria, spread out evenly across the plate, and feed essentially as "loners." They tend to move away from one another if they collide. In other strains, the worms always group together in visible clumps, where they feed as "socializers." They actively brush up against one another, squirming and winding around one another as they feed. This behavior is induced by the presence of food; when food is absent, the socializers also spread more or less evenly across the plate (Fig. 4.3).

It turns out this difference in feeding behavior is determined by which of two alleles of a single gene, called *npr-1*, is present in a given strain. The *npr-1.215f* allele is found exclusively in social strains, whereas the *npr-1.215v* allele defines the loner strains. It is important to note that these are natural alleles occurring within the species, and not induced in the laboratory. When the *npr-1.215v* form of the gene was used to replace the *npr-1* gene in a social strain of worms, they became solitary feeders (see Fig. 4.3). It turns out that the *npr-1* gene encodes a protein that is very similar to a neurotransmitter receptor in mammals called "neuropeptide Y." The two alleles produce forms of this molecule that differ by only a single amino acid out of several hun-

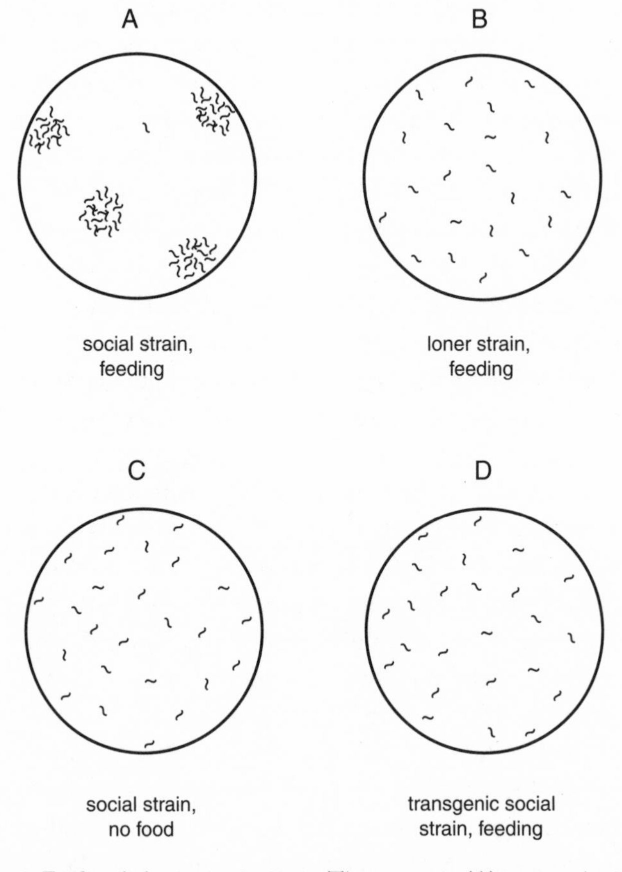

Figure 4.3 Feeding behavior in *C. elegans*. The worms in (A) express the 215f allele of the npr-1 gene; "loner" worms (B) express the 215v allele. The clustering behavior in (A) is food-dependent (C). When the 215f allele in "social" worms is replaced with the 215v allele, the worms exhibit loner behavior (D).

dred making up the protein. *C. elegans* does not use neuropeptide Y as a neurotransmitter, but produces molecules similar enough to neuropeptide Y that they could interact with the *npr-1*-encoded receptor. Apparently the "social" worms use this system to communicate with one another during feeding.

One of the most important behaviors, and one not seen in evolution prior to the appearance of multicellular life forms, is associative learning, or, as it has been called since Pavlov's time, "classical conditioning." This type of learning requires stable, retrievable memory of previous experience, as well as the ability to respond quickly to an environmental cue.

A recently published experiment suggests that classical conditioning and associative learning have already made their appearance at the stage of evolution represented by *C. elegans*. These worms have separate sensory neurons that allow them to detect sodium ions (Na^+) and chloride ions (Cl^-) in their environment, and they will normally move toward moderate concentrations of either ion. In an initial conditioning step, a food source that appeals to worms (bacteria) or a substance they strongly dislike (garlic extract) was mixed in water with either ion. The worms were then immersed in the various resulting mixtures, after which they were washed thoroughly and placed on a dish. A drop of water containing either Na^+ or Cl^- (but neither food nor garlic extract) was placed nearby. If the worms had previously been exposed to bacteria in association with the ion present in the water droplet, they entered it and actively explored for food. If the worms had previously been exposed to garlic extract in association with the ion present in the water, they avoided the water droplet. In another set of experiments by the same researchers, the same kind of associative learning was established between smell and a food source.

The worms retained memory of their conditioning experience for many hours, and used that memory to guide their behavior in several subsequent tests. This in itself is an extraordinary finding. The behavior elicited in these extremely simple worms has all of the hallmarks of the learning behavior expressed in animals with much more complex nervous systems. They were able to associate a positive (bacteria) or negative (garlic) reward with a totally independent second signal (Na^+ or Cl-), and move toward or away from the test droplet accordingly, even though the positive or negative reward was nowhere around.

Moreover, in the development of this memory, the researchers were able to distinguish a period of short-term memory, followed by establishment of long-term memory, similar to processes we ourselves use.

But the researchers made an even more astonishing discovery: single-gene mutations could abolish associative learning in these worms, without noticeably affecting other biological functions. By introducing random mutations into the genomes of their subjects with a mutagenic chemical, and screening the treated worms for their ability to learn in the conditioning experiment just described, they found two genes which, when mutated, abolish the ability to learn. These two genes have been named "*learn-1*" (*lrn-1*) and "*learn-2*" (*lrn-2*).* Worms with mutations in either of these genes are still able to detect Na^+ and Cl^-, and they still love bacteria and hate garlic. They are perfectly able to move around, and will respond appropriately to food sources and to ions. But they are incapable of learning to associate the presence of a particular ion with a particular food source.

The *lrn-1* and *2* genes have not yet been isolated and cloned, so we do not yet know exactly what they do. Like the vast majority of genes, they most likely encode proteins, and these proteins doubtless play an important role in associative learning. It should not take very long to determine that role. And this is where one of the major advantages of *C. elegans* as an experimental system—for the extensive mapping of its neurons—will come into play. The first step will be to find out which neurons are involved in this new associative learning test. That will be done by using microlasers to kill off the neurons one by one, until those found to be crucial in associative learning are identified. Clearly, the sensory neurons used to detect food and ions will be involved; anything that compromises their function will inhibit learning. Any motor neuron that controls the ability of the worms to move toward or away from a food source could also be involved. But all of these responses were analyzed in the experiments just described, and all are fully operational in the *learn-1* and *2* mutants. So it is most likely the interneurons, and probably those in the neck ring, that will be of greatest interest.

The second step will be to isolate, clone, and sequence the *lrn-1* and

* When geneticists do not know what a gene codes for, but only the mutation it causes, they usually assign the gene an abbreviated, italicized named based on a contraction of the name of the mutant assigned by its discoverers.

-2 genes themselves, so their function can be established. The process of isolating a completely unknown gene can be long and tedious, but it may not have to be in this case. The *C. elegans* Genome Project was recently completed, and has resulted in the cloning and sequencing of each of the approximately 13,000 genes *C. elegans* possesses.* It should not take long to match up the *learn* mutations with the corresponding *C. elegans* genes. We will then be able to study the allelic changes in the mutant forms of these genes, and try to understand how these allelic alterations underlie the behavioral variability we see. The combination of precise knowledge of which neurons are involved, and eventual knowledge about the function of the *lrn-1* and *-2* genes and their alleles, will allow a level of analysis of the genetic control of behavioral function never before attained.

Because of the tremendous evolutionary conservation of the elements underlying nerve cell function, it is very likely that human counterparts of *lrn-1* and *-2*, if they exist, will also be found in the near future. A fairly complete "library" of genes expressed in the human brain has already been established, but the function of very few of them is actually known. The *C. elegans* genes can be used to screen this library for their human homologs,** and the function for these genes we find in *C. elegans* will likely be the function they serve in the human nervous system as well. We can then ask whether variants of these genes among humans correlate with variations we see in a range of human behaviors, particularly learning and memory. Thus, research on a lowly roundworm has the potential for providing important insights into human behavior, and the functioning of the human brain.

---

* Sequencing of the human genome, with its estimated 100,000 genes, is scheduled for completion in about the year 2003.

** The techniques for isolating, cloning, and sequencing genes, and for the use of isolated genes to screen gene libraries, is described in W. Clark, *The New Healers: Molecular Medicine in the Twenty-first Century* (New York: Oxford University Press, 1997.)

5

About Genes and Behavior

In the preceding chapters we have presented a good deal of evidence, both direct and indirect, for a role of genes in explaining a substantial portion of the variability we see in the behavior of both animals and humans. But we have said very little about *how* genes might do this. In part, that is because we do not yet know all the details of how genes influence behavior. As we see below, there is good evidence that most behaviors, especially in humans, are genetically complex; that is, they are influenced by not just one gene, but by many. But until we know what the individual genes influencing a behavior are, it is difficult to say anything about how they may act as a group, and we are still at the stage of identifying the individual genes.

It is easy for mutations in a single gene to disrupt a behavior, as we saw in the case of the *lrn-1* and *-2* genes in *C. elegans*. There are also numerous examples of individual defective genes disrupting human behavior, such as the genes underlying Huntington's disease and the heritable forms of Alzheimer's disease. But that does not mean that the behavior in question is controlled by a single gene. Any behavior is always an outward expression—a phenotype—reflecting the operation of biological systems regulated by many different genes, interacting with each other and with the environment. The numbers of genes involved in most behaviors would certainly be in the range of dozens, and perhaps even hundreds. That mutation of a single gene can often disrupt an entire behavior simply reflects the fact that the biological systems underlying that behavior are tightly regulated; breakdown of a single component can often shut down an entire pathway. Think of all the individual parts in a computer required to store and retrieve a document from memory. If one of the components involved should break down, memory function is lost. But that certainly does not mean that memory in the computer can be accounted for by the single part that broke. So it is with genes involved in determining complex human traits.

The underlying issue of complexity in the genetic regulation of behavior will be very much tied up with the types of genes we believe will be involved in that regulation. What kinds of genes will we be looking for as we begin to dissect the genetic basis of behavior? When we think of a particular human behavioral trait—boldness versus shyness, for example, or curiosity versus indifference—should we expect to find unique genes whose alleles directly and specifically cause the varying phenotypes we see in these traits in the individuals around us?

In thinking about the kinds of genes we might expect to see influencing various human behaviors, it might be useful to take a brief look at two other areas of human biology where scientists have gained important mechanistic insights by ferreting out the genes involved: cancer and aging.

For most of its clinical history, cancer was viewed as a thousand or more different diseases—at least as many different cancers as there were different cell types in the body, each requiring different treatment, each with its own outcome. Although there were hints that at least some kinds of cancer might have a genetic basis, no one knew where to begin. It seemed that if cancer was indeed caused by genes, then there would have to be an enormous number of different cancer genes—at least one gene for each different cancer type. For many years that possibility inhibited scientists from trying to unravel the genetic basis of cancer.

Fortunately, it didn't discourage all scientists from pursuing this question. And those who stayed in the game ultimately found out that there really are no "cancer genes" per se. In fact, there are relatively few genes involved in causing cancer, and we find the very same genes—or at least the same classes of genes—underlying each cancer, regardless of tissue origin. Cancer turns out to be caused by allelic variants of the genes responsible for regulating normal cell division. Cancer cells, at least when they first arise in the body, are no different from their noncancerous neighbors, with the exception that they have somehow acquired allelic variants of cell division genes that result in cell division when there should be none. Otherwise, the cells are perfectly normal. Different cancers are different because the cells in which they arise are different, but the genes responsible are largely the same. This new mind-set about what cancer represents has revolutionized strategies for both the detection and the treatment of cancer, strategies that are already finding their way to the clinic.

A similar shift in thinking has occurred in the study of aging. When we add up all of the outward manifestations of aging, and combine them with all of the internal age-related changes detectable by laboratory tests, we end up with a staggeringly long list of degenerative alterations associated with age. It seemed to some researchers that an enormous number of different processes must be taking place during aging in the various tissues and cells of the body to explain such a diversity of measurable events. And again there was the assumption that to the extent that aging is genetically controlled, there must be enormous numbers of genes involved—at least as many different genes as there are different aging phenotypes.

Yet this turns out not to be the case. A degenerative process involving alterations in the activity of a single gene, expressed in all cells in the body, can compromise the function of many different cells in the body, but each in a different way. For example, in Werner's syndrome, a mutation in a single gene causes accelerated aging in many different tissues. Beginning at age twenty or so, individuals with Werner's syndrome develop gray hair, aged skin, bone loss, muscle wasting, cataracts, and cardiovascular disease, among other things. And all because of an alteration in a single gene, in this case a gene involved in unwinding DNA—perhaps in preparation for repair of the DNA damage that is thought to be a major contributor to aging. The

Werner's gene is by no means the only gene involved in aging—we surely will find others—but the rather startling range of phenotypes caused by this single gene, across such a large number of cell types, suggests the possibility that a relatively small number of genes could account for a large proportion of aging phenotypes.

Cancer and aging are not behaviors, but a study of their underlying genetics has important implications as we think about behavioral genetics. First, as we have seen, just because a particular biological phenomenon is complex does not mean a priori that large numbers of genes are involved. Certainly, cancer and aging involve more than single genes. But each type of cancer, or each aging phenotype, does not necessarily involve separate and distinct sets of genes. Second, the genes involved in modulating a particular phenotype may or may not have an obvious connection with the phenotype. Although in retrospect we might have made the connection several decades ago between genes controlling cell division and cancer, we might never have connected cataracts in the elderly with a DNA repair enzyme, if someone had not discovered the gene first and shown its involvement in Werner's syndrome.

DNA and the Language of Genes

There are an estimated 50,000 to 100,000 genes spread out among the twenty-four chromosomes that make up the human genome: twenty-two "autosomes," present in each cell as pairs, plus the X and Y sex chromosomes. The genes are contained in long, linear strands of DNA stored in the nucleus of each cell in the body. Humans have an enormous amount of DNA—about a meter of it per cell. If all of the DNA in an adult human being (10^{14} cells, give or take a few trillion) were strung out end-to-end, it would reach from the earth to the moon and back many thousands of times.

DNA is made from small chemical units called nucleotides (Fig. 5.1). There are just four of these units, called by their abbreviations A, C, G, and T. They are linked side by side into individual DNA strands, with no restrictions on linear sequence; any nucleotide can lie next to, and hook up with, any other nucleotide along the same strand. In cells, DNA always occurs in the form of the well-known "double-helix"—two individual strands wound about one another to form a helix. The

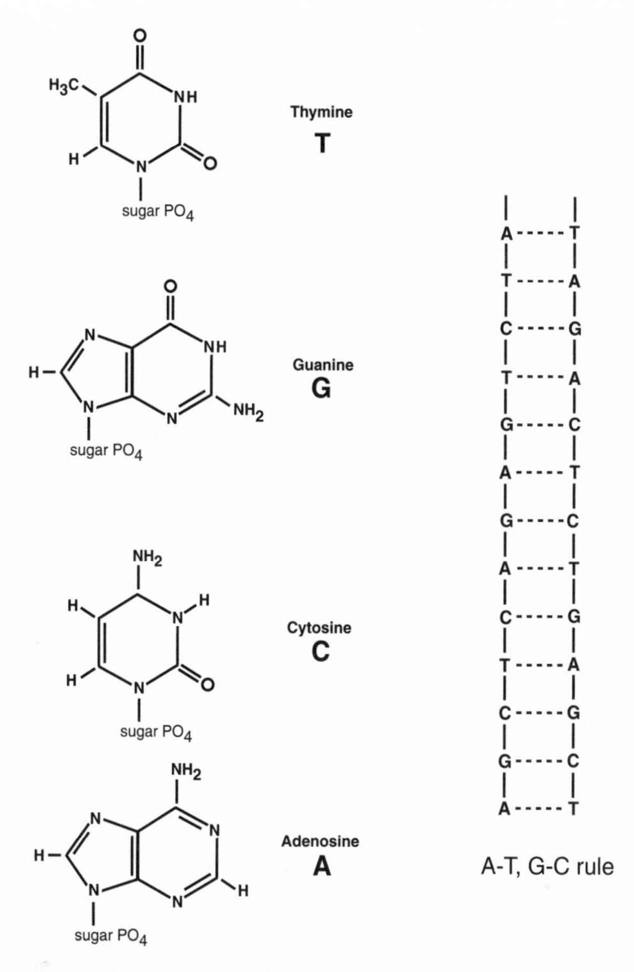

Figure 5.1 The four nucleotide subunits used to make DNA.

nucleotides facing each other across the two strands of the double helix form a sort of bond with one another to stabilize the helix, and here there are restrictions on who can shake hands with whom. An "A" can face only a "T" across a double helix, and a "C" can lie across from only a "G" (Fig. 5.2).

This restriction on pairing between nucleotides in opposing strands is the secret to faithful DNA replication. When a cell divides, the two strands must each replicate, so that an identical DNA double helix can be transmitted to each of the new daughter cells. As can be seen in Figure 5.2, when two DNA strands pull apart, each serves as a template for assembly of a new strand; new copies of individual nucleotides

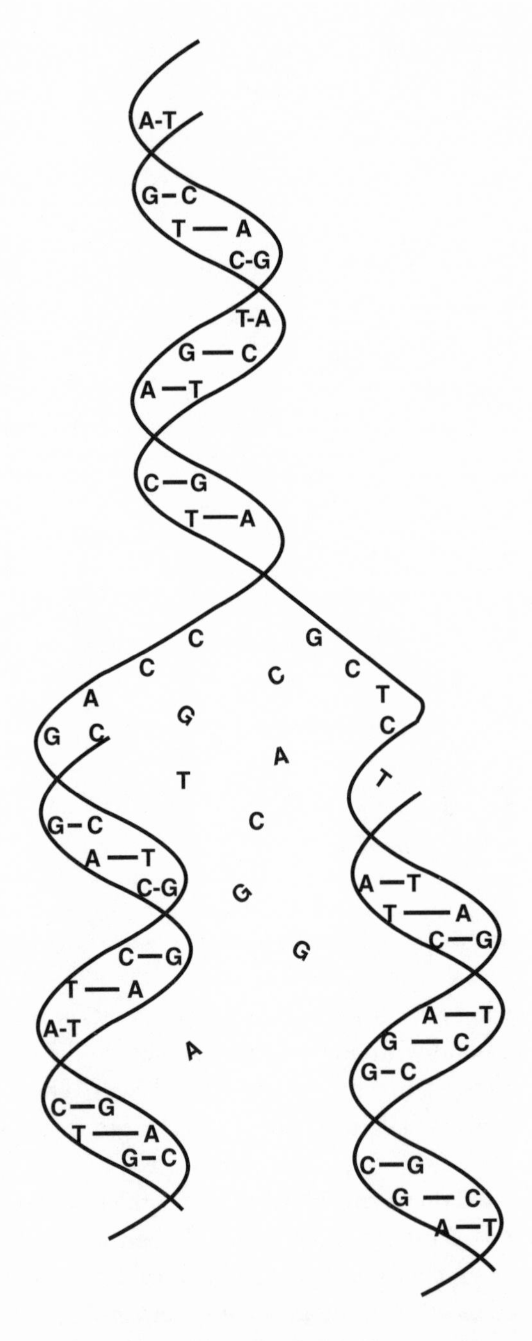

Figure 5.2 Detailed structure of replicating DNA. New strands are generated from free-floating nucleotides, using the unwound strands as a template.

are brought in and lined up along the template strand according to the pairing rules, A with T, and G with C. Once the new set of nucleotides is lined up along the template, and locked into place with each other and with the template strand, voila! We have two new double helices that are exact replicas of the original helix. This insight, among others, earned a Nobel prize for James Watson, Francis Crick, and Maurice Wilkins in 1962.

Genes are defined stretches of nucleotides lying along one or the other of the two strands of DNA (both are used), marked by a starting point and a stopping point. The nucleotides between these two points are read in groups of three, with each such triplet specifying a particular amino acid; the gene as a whole thus defines a particular protein. Alleles of a given gene represent slight variations between individuals in the nucleotide sequence of that gene, which results in minor amino acid variations in the corresponding protein. Alleles arise because the DNA copying process is not always perfect; when DNA is replicated at the time of cell division, so-called copy errors creep into genes from time to time. If the new variant is kept, it becomes an allele of the gene from which it arose.

The way in which genes give rise to proteins within a cell is shown in Figure 5.3. A given gene is first copied into a form called "messenger RNA." RNA is very similar to DNA. The advantage of making separate RNA forms of genes is that individual gene copies can be moved out of the nucleus and into the portion of the cell where proteins are made, without disturbing the rest of the genome. Once at this site, the messenger copy of the gene is attached to a small structure called a ribosome, where its sequence is read off, triplet by triplet, and converted into the amino acid sequence of a protein.

Each gene consists on average of about a thousand nucleotides. Even at the upper limit of 100,000 genes in the human genome, it is obvious that genes account for only a very small portion of all the DNA we carry around in our cells. It turns out that actual genes are scattered rather widely throughout the genome, separated by vast stretches of DNA that do not code for anything (Fig. 5.4). Even within genes there are stretches of nucleotides, called "introns," that don't code for anything. What does all of this extra DNA do? Some of it surely represents genes we once used, far back in evolutionary time, and no longer need. Some of it probably serves as raw material for gen-

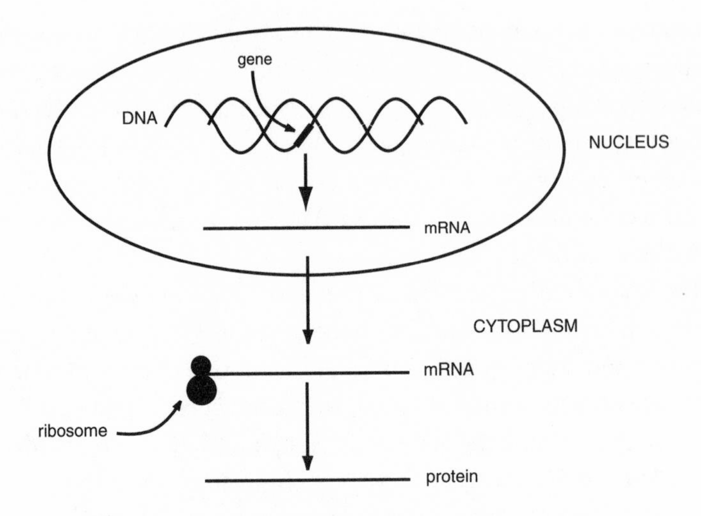

Figure 5.3 Information flow: DNA RNA protein.

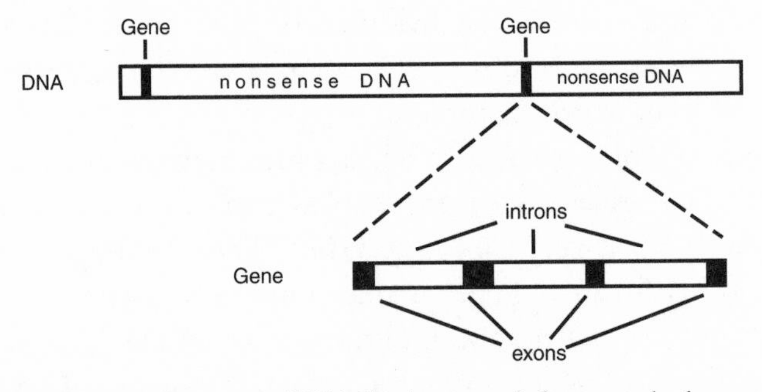

Figure 5.4 Sense and nonsense DNA. In the expanded gene, only the portions in black (exons) actually code for the protein corresponding to the gene.

erating new genes. The bottom line is we simply do not know. It is often called "junk" DNA, or "nonsense" DNA. But as discussed in later chapters and in Appendix I, below, some of this DNA has properties that make it very useful for "tagging" selected regions of DNA, and following them around from one generation to the next.

The One and the Many: Single-Gene versus Multigene Traits

Most of the genetics that was done in the twentieth century, both classical transmission genetics and even the more recent molecular genetics, has involved the study of single genes—how they are inherited and which properties of the individual organism they affect (transmission genetics), and the structure, function, and regulation of the underlying gene (molecular genetics). Mendel, for example, studied inheritance in peas. He proved that certain traits, such as the texture or color of peas, or the height of plants, were controlled by individual "units of inheritance" that eventually came to be known as genes. Heredity had previously been viewed as some sort of "blending" of the overall characteristics of each parent. The idea that individuals are composed of a large number of individual traits that can be passed forward separately and distinctly from one another was a major revolution in the way we view heredity.

Mendel was also the first to describe alleles of genes. Alleles are slightly different forms of the same gene present within a given population. These differences arise through the various processes of mutation, which result in nucleotide changes in the sequence of a gene. The resulting changes in the protein encoded by the altered gene (the allele) can be harmful, neutral, or favorable with respect to the function of that protein. Harmful mutations generally do not survive the winnowing processes of natural selection. Neutral or favorable changes are usually kept. Favorable alleles, over time, can increase their frequency substantially within a population if they result in improved reproductive efficiency. Mutations can arise because of physical damage, such as that caused by radiation and certain chemicals that damage DNA. But perhaps the major source of mutations that alter genes within a population is the introduction of copy errors during the normal reproduction of DNA in dividing germ cells. Radiation and chemical damage is fairly easy to detect and repair, but copy errors are much more subtle, and often result in uncorrected changes in genes, thus giving rise to new alleles for natural selection to act upon.

A majority of the genes making up the genomes of most species have allelic forms; such genes are referred to as "polymorphic," or having many forms. In terms of the genetic basis of variable behaviors, we are concerned primarily with polymorphic genes. The question behavioral geneticists ask is to what extent the variability we observe in behavior within a population can be ascribed to genetic differences. Genetically based differences arise principally through the inheritance of different alleles of polymorphic genes. So throughout the remain-

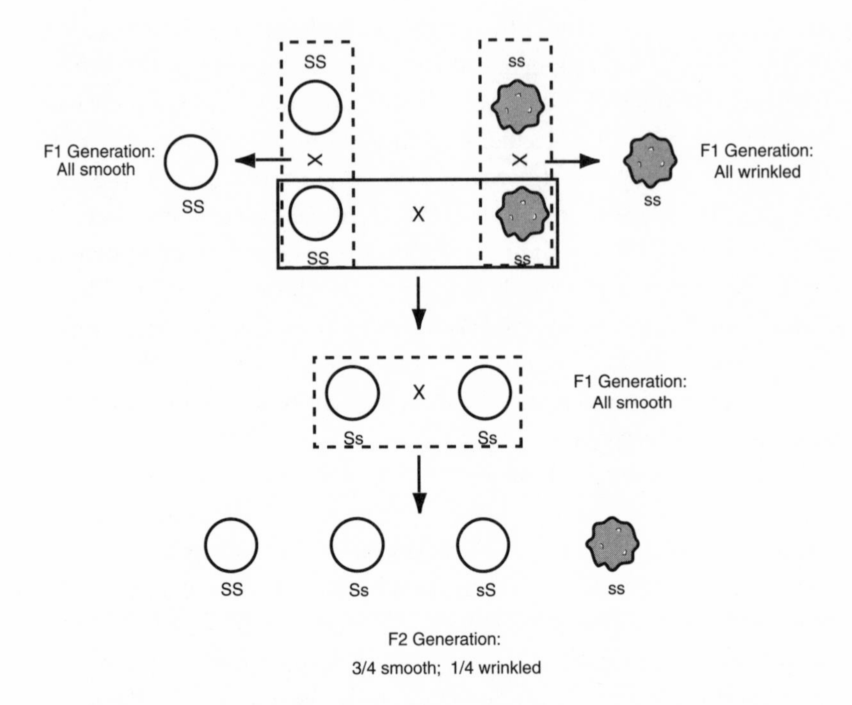

Figure 5.5 Mendel's results from a pea breeding experiment. Breeding pairs are enclosed within boxes.

der of this book, we are not talking so much about genes that regulate behavior, but rather different alleles of individual genes that contribute to differences in behavior.

Mendel also defined the important concept of dominant and recessive alleles. Each individual has two copies of each gene, one inherited from each parent. If there is more than one allele of a given gene within the population, an individual may have two different alleles for a given gene, in which case that individual is said to be heterozygous for that gene. The protein each allele encodes usually has slightly different properties. For example, Mendel identified a gene governing the quality of the skin covering individual peas (Fig. 5.5). One allele of this gene (S) results in smooth skin, and the other allele (s) codes for wrinkled skin. Mendel showed in his breeding experiments that if an individual plant had two "smooth" alleles (in genetic terms, *homozygous* for the smooth allele, i.e., SS), the peas of that plant would be smooth. If the individual had two "wrinkled" alleles (homozygous wrinkled, ss), its peas would be wrinkled. If a plant had one wrinkled and one smooth allele (*heterozygous*, i.e., Ss), the resulting peas were not halfway between wrinkled and smooth; the traits did not blend. The peas in such a plant were all perfectly smooth. The smooth allele is therefore said to be dominant over wrinkled. The important thing to recognize is that there is no observable (phenotypic) difference between the peas of plants that are homozygous smooth (SS) and heterozygous (Ss).

A major question confronting behavioral geneticists is to what extent classical Mendelian (single-gene) inheritance is useful in explaining the influence of genes on behavior. There are a few examples of single genes governing a behavior, even in humans. For example, there is a single gene that determines whether individuals can taste the chemical phenylthiocarbamide (PTC). Genes that affect the ability to taste definitely affect behavior, in the sense that they are involved in an organism's response to the environment. The PTC taste gene in humans has two alleles, one conferring the ability to taste PTC, and one that results in an inability to taste PTC. The allele conferring the ability to taste is dominant: If an individual has one copy of each allele, he/she can taste PTC. In humans, we can see that this trait is governed by a single gene by looking at inheritance patterns in families. Roughly 25 percent of the population cannot taste PTC, which indicates that the recessive allele of this gene is present in about half the population.*

In animals, where we can do controlled breeding experiments, the difference in inheritance patterns can be determined in a slightly different way. Imagine that a *ptc* gene exists in mice, with the same dominant/recessive pattern as in humans. (For simplicity's sake, we assume the existence of only a single dominant allele and a single recessive allele.) Imagine further that we begin randomly breeding mice, and testing the offspring for the ability to taste PTC. We then begin selectively breeding together those offspring who, on the one hand, cannot taste PTC, and separately those offspring who can taste PTC in their food.

---

* We do not know the significance of the ability to taste PTC, or the reason for the existence of two distinct alleles controlling its taste in humans. PTC is not found in food; our ability to taste it presumably reflects a random cross-reactivity of PTC with some standard taste receptor. The alleles we are discussing here would be alleles of that receptor.

Possible results of such a breeding experiment are presented in Figure 5.6. Within essentially a single generation, we would have a subline of mice all of whom were unable to taste the chemical (curve A). We would be selecting for mice with two defective (recessive) copies of the *ptc* gene; two such mice mated together would have no good copies of the gene to pass on. They would thus transmit this deficiency to 100 percent of their offspring each time they mate. It would take a somewhat longer number of generations to derive mice all of whom were able to taste PTC (curve B in Fig. 5.6). This reflects the fact that phenotypically, there is no difference between a heterozygote carrying one dominant and one recessive gene, and two dominant genes; both can taste PTC. The recessive gene will disappear only gradually from the population, whenever two copies show up in the same individual (who is then eliminated from the breeding pool). If the ability to taste PTC were controlled by multiple genes, each with dominant and negative alleles, the ability to generate either type of subline would look like curve C.

Let's look at the breeding and selection pattern for a genetically complex personality phenotype in mice—fearfulness versus adventure

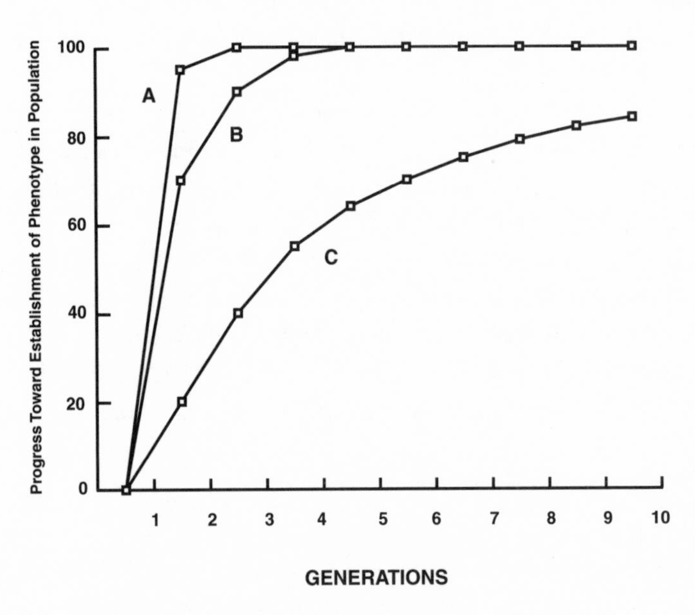

Figure 5.6 Selective breeding for the ability to taste PTC.

seeking. If individual field mice are placed in a large, open box, under full light, we see a range of behaviors displayed by individuals in an unselected population. Some mice will become almost catatonically fearful; they will cringe in one corner, and not move. They will urinate and defecate. The majority of mice will show a little more curiosity, and explore their surroundings somewhat, some more boldly than others, but generally hugging the walls as they move about. A few mice will roam around the entire floor surface of the box, exploring, seemingly unconcerned about anything at all.

There could well be implications of this range of behaviors in a population of mice for survival in an open field. In a time of plentiful food, we can imagine that fearful mice will have a survival advantage because they are less likely to be eaten by a predator than are their more adventurous relatives. Caution would be a good thing in that case. But when food is scarce, the fearful mice may be at a disadvantage because they will be too timid to seek out resources aggressively. They might also be less successful in finding mates. So having a range of behavioral traits related to fearfulness could be advantageous for the species as a whole.

When we try to develop sublines of fearful mice and bold mice, starting with a normal, unselected population showing all of the variants just described, the result is very different from what we saw with the ability to taste PTC, which is controlled by a single gene. With video cameras and electronic detectors, mice placed in an "open-field box" of the type described above can be monitored precisely for the extent to which they move about and explore their surroundings. The mice are allowed to breed, and the offspring are scored in an open-field box and segregated into most fearful and least fearful subgroups. The procedure is then repeated, but with subsequent breedings allowed only within each subgroup. After each mating, the offspring are again sorted into most fearful and most adventurous subgroups, and the inbreeding continues.

The results of such a breeding program do in fact look like curve C in Figure 5.6. After perhaps a dozen generations of breeding and selection for the desired traits, sublines of mice will have been generated that differ by a factor of a hundred or more in fearfulness. Mice from the fearful subline freeze up almost completely when placed in an open-field box; the bold mice move about and explore almost without fear. It is important that they do this whether they were reared by their own mothers, or were transferred immediately after birth to a surrogate mother of the opposite behavioral phenotype for nursing and rearing. When newborns from each strain are mixed and placed with a surrogate mother, each will go on to express its own genetic predisposition, rather than that of its surrogate littermates or its surrogate mother. This experiment tells us several important things. First, since these traits are passed on faithfully to offspring independently of contact with others of their species, we can conclude that they are inherited and not learned. But the degree to which the genetic predisposition to fearfulness or fearlessness is expressed continues to change across a large number of selective breeding generations. The most likely interpretation of this is that there are many different genes involved in regulating these behaviors, and that a large portion of these genes have multiple alleles affecting the degree to which each of them contributes to the overall phenotype. It is unlikely that we will find a single gene that we can call "*frfl*" (for "fearful") that completely dominates this phenotype.

From the fact that boldness as opposed to fearfulness is heritable and not learned, we know it is largely a genetic trait. But it is not the case that all individuals in a natural (unselected) population are either fearful or fearless. If scores for individuals are plotted on a common graph, there is a continuous range of phenotypes from extremely fearful to very bold, with most individuals lying somewhere in between in a classical bell-shaped curve (Fig. 5.7). The individuals at either end of the curve approximate the members of the two sublines we generated by selective breeding: very fearful or extremely fearless. This range of genetically controlled phenotypes within an unselected population is the definition of what is called a "quantitative genetic trait." A number of different genes control the trait, and the position of any member of an unselected population on this curve is a reflection of that individual's particular collection of alleles for the various genes underlying the fearfulness-versus-bold trait. Virtually every behavior that has been analyzed in human populations generates a curve such as that shown in Figure 5.7.

While behavior in an open-field test is clearly a quantitative trait, we can discern the effects of individual genes in the overall behavioral pattern. Albinism in mice is known to be controlled by a single recessive gene. There are mice that are genetically identical except for having different alleles of the albinism gene. The albino mice are much more tentative than their nonalbino littermates. This is mostly because the albino mice are much more sensitive to bright, full-spectrum light. In a field test carried out with red light, to which the albino mice are much less sensitive, they score about the same as non-albino mice in terms of moving about in an open field. These kinds of subtle interactions involving many different types of genes underscores the complexity of sorting out the genetic contribution to many behaviors.

Essentially all of the animal and human behaviors we discuss in the remainder of this book are quantitative traits. One of the major goals of molecular biology and molecular medicine in the coming decades is to identify as many genes associated with quantitative traits as we possibly can, whether these be traits for behavior, disease, or any other complex aspect of human biology. It might seem to make sense that the genes associated with any given quantitative trait should all be clustered next to each other on a single chromosome, where they could interact with one another. That is not the way it works, however; the genes involved with a single trait are usually spread more or less randomly throughout the entire genome. The chromosomal locations of the various genes associated with a given quantitative trait, like the albinism gene that affects mouse behavior in an open-field test, are

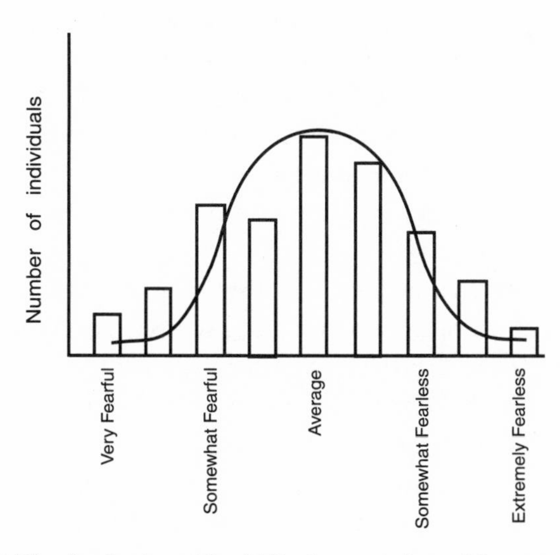

Figure 5.7 The distribution of fearful/fearless types in a wild mouse population forms a typical bell-shaped curve.

called "quantitative trait loci," or QTLs. Whereas some of the genes at these QTLs may interact with one another in a highly interdependent fashion in creating the ultimate phenotype characterizing a particular behavior, other genes may affect that behavior relatively independently of one another. Most important, not all QTLs will affect a given trait equally; some will be "heavy hitters," accounting for a quarter or a third of the genetic component of that trait. Others will make relatively minor (but, by definition, detectable) contributions to that trait. Molecular biologists tend to be interested in all of the identifiable QTLs underlying a trait; molecular medicine is generally concerned with the heavy hitters.

But genes cannot explain all of the variability in quantitative traits; environment, too, has a role to play. This is a point often overlooked when interpreting the results of studies in behavioral genetics. Most data consistently show an effect of the environment ranging from 30 to 70 percent for different behaviors. There are two important points to bear in mind about the role of environment in determining behaviors. First, when we measure behavior we are not measuring genotype, we are measuring phenotype. Phenotype is determined *only* by the interaction of a particular genotype with the environment. The second point is that two identical genotypes, placed in two different environments, may produce two quite different phenotypes with respect to any particular characteristic, behavioral or otherwise. In many of the studies we discuss in this book, involving monozygotic twins with identical genotypes, the impact of the environment on many behaviors often seems rather minimal. That is largely because the different environments we are talking about are not always that different, and the twins involved can manipulate and extract from their respective environments quite similar things, in the end.

The impact of environment on behavior can be shown quite clearly in a variation of the open-field test just discussed. Newborn rabbits are somewhat unique in that the mother spends very little time with them. She is away from the nest almost the entire day, and nurses them only once per day, usually in the evening after she returns from foraging. Even rabbits maintained in a laboratory environment follow this pattern of minimal contact with newborn young. When recently weaned rabbits are placed in an open-field test box, they exhibit the same range of responses seen in rats and mice: Some huddle, almost immobilized, against the walls, while others explore the field box with varying degrees of fearlessness.

Researchers found that by subjecting young rabbits to a variety of sensory experiences during the preweaning period—handling them or exposing them to mild shocks or higher or lower than normal temperatures—they could greatly increase the tendency of these animals to explore their surroundings in an open-field test later on. Increases in the degree of adventurousness as adults correlated directly with the amount of preweaning handling. But unlike the degree of gene-biased fearlessness produced in selectively bred mice, the fearlessness developing as a result of early life environmental experience is not passed from one generation to the next. The offspring of two "environmentally generated" fearless parents will show the same random assortment of fearful/fearless phenotypes as the offspring of two timid parents.

Environmental effects can also be seen in standard inbred strains of mice. These strains are created by starting with a single pair of mice, and repeatedly breeding only brother and sister descendants in each generation. The object here is not to select for a particular trait, but just to achieve strains of mice in which all members are in effect the equivalent of human identical twins, which is achieved after about thirty generations. These kinds of strains have proved invaluable for studying cancer, organ transplantation, and other physiological phenomena where precise genetic definition is critical.

All of the members of a given inbred strain are genetically identical, but each inbred strain is different from every other strain. Differences between strains for phenotypes such as fearfulness/boldness, although not selected for, do show up, and are presumed to be largely genetic. But differences among members of the same strain cannot be genetic, by definition, and we do see modest differences in such traits among members of the same strain. These differences cannot be enhanced by selective breeding, because they are not transmitted from one generation to the next. How such differences arise in mice is not entirely clear. While most colonies are managed in a highly uniform manner, differences may occur in the way individual mice are handled by animal care technicians. Different degrees of crowding during the early part of life, different ratios of males to females in the same cage, different experiences with viral or bacterial pathogens could well affect behavior. It is never possible to control environmental factors completely, and the variabilities we see within inbred mouse strains makes it absolutely clear that these factors can affect behavior. That such differences are not learned is shown by experiments wherein, for example, newborn pups from a fearful mother are transferred at birth to an aggressive foster mother. As adults, the variability among these mice in fearfulness is not significantly different from mice reared by fearful mothers.

A clear example of the interaction of genes with the environment in humans is the drastic increase in obesity among individuals in industrialized countries in the past hundred years. Most studies suggest that the frequency of individuals who are overweight has almost quadrupled since the end of the nineteenth century. There is no way that the genetic composition—the distribution of alleles affecting body weight—of any population breeding as slowly as humans could have changed in so short a time. The change must be primarily environmental: Changes in diet and much less expenditure of energy in the daily tasks of living are almost certainly the culprits. These environmental changes have interacted with the existing pool of individual genotypes in different ways, and many of the underlying changes in body weight are purely behavioral in nature. We look at this question in more detail in Chapter 10.

Going for the Genes

One of the goals of behavioral genetics is to find and identify the genes in which allelic variation could be responsible for the variability we observe among individuals for a given behavior. We are thus looking only for polymorphic genes, genes present in multiple forms among the members of a given population of animals or people. Since behavior is for all practical purposes controlled by the brain and nervous system, with help from certain hormone-producing tissues, we will be looking at genes expressed in those systems. But how do we go about finding those genes?

The most important tool in the isolation and identification of any gene is a good genome map. A genome map consists of physical pictures of the individual chromosomes making up the genome, with the location of what are called DNA markers clearly indicated. In the early days of genome mapping, DNA markers were simply genes whose location had been very laboriously traced to specific chromosomal locations. In research animals such as *C. elegans* or fruit flies, the presence of individual genes is usually first detected by mutations giving rise to observable phenotypes. Once it has been established that a given gene is physically associated with a particular chromosome (the first one is always the hardest!), that gene can serve as a "marker" for that chromosome. When a new mutation appears, we can ask whether or not this mutation is inherited along with alleles of the marker gene. If it is, then it must be on the same chromosome as the marker gene. Additional genes traced to this same chromosome can then be mapped even more precisely.* Starting early in this century, researchers have gradually identified the chromosomal locations of a large number of genes that are polymorphic within fruit fly populations.

It has been much more difficult and painstaking to establish the locations of genes in the human genome. In fact the first gene was not mapped with a specific chromosome until 1968. Humans have twenty-three chromosomal pairs, compared to four for fruit flies, and considerably more genes. Nevertheless, by the 1980s fairly precise chromosomal locations for a thousand or so human genes had been established. Today, as part of the Human Genome Project, markers are being established on each and every chromosome to aid in gene mapping.

The locations of genes already mapped serve as ready markers, but molecular geneticists have moved well beyond that. They have established DNA markers that have nothing to do with genes per se. True genes are not lined up end-to-end continuously along a chromosome; they are separated by stretches of DNA that do not code for anything, or "nonsense" DNA. Nevertheless, scientists have found that some portions of this DNA are remarkably well conserved from generation to generation, and are even present in the population (and inherited faithfully) in allelic forms. These nonsense DNA markers can be used in the same way as allelic genes to pinpoint the chromosomal location of new genes. The section of the DNA containing that marker can then be isolated and systematically searched for the new gene. Once the gene is isolated, it can be sequenced, and the nature of the protein

* The way in which DNA markers are used to determine the location of a new gene is described in more detail in Appendix I.

it encodes can be inferred from the DNA sequence. And of course, that gene itself now becomes a marker on the map of the chromosome on which it was found.

The availability of a detailed set of DNA markers for the human genome (the first sets have become available only in the past few years, and are still being refined) provides molecular geneticists with an alternate approach to identifying the multiple genes associated with complex, quantitative genetic traits. This extraordinarily powerful new tool is called a "genome scan." For a given quantitative trait, one can correlate inheritance of that trait with inheritance of the full range, or with selected subsets, of human genome DNA markers. This requires large numbers of individuals, preferably representing at least three generations. This technique can tell us which DNA markers are inherited along with a given trait, thus revealing the approximate chromosomal location of the underlying genes. The sections surrounding the DNA marker can then be examined for the presence of candidate genes. The great advantage of a genome scan is that it provides a systematic means of uncovering trait-associated genes whose existence was previously unsuspected.*

From what we have discussed in this chapter, there are several things to carry with us as we continue to explore the genetic basis of behavior. It is true that most behaviors are quantitative genetic traits—they are almost certainly not going to be governed by one or even a small number of genes. And the effects of these genes will, to varying degrees, be subject to modification by the environment. But from what we understand of other phenomena, such as cancer and aging, it may well be the case that the genes influencing behavior will not be very different among different behaviors. So should we set out to look for "fearful" genes? Should we expect to find genes dedicated only to mental functioning? Will special genes guide us to become artists, or engineers? A few decades ago, many scientists might have said yes. Today, they are hedging their bets, but now they have powerful new tools to reshape their answers. In the next few chapters we examine some of the experiments that are causing researchers to take a more cautious, yet hopeful, approach to understanding genes and behavior.

---

* Details of how a genome scan is used to detect previously unsuspected genes are given in Appendix I.

6

Life in the Fourth Dimension

The Role of Clocks in Regulating Behavior

One of the most constant features of the environment in which biological organisms evolved, and in which they live today, is periodicity: the marking off of the days, the months, and the year; the ebb and flow of the tides and the seasons. Other features of the earth—the composition of its seas and its atmosphere, and the amount of thermal and electromagnetic energy at its surface—have changed significantly since life first appeared. But not its geological rhythms. When life first appeared several billion years ago, it emerged and evolved in the context of these rhythms, which became embedded in the very fabric of life itself in the form of biological clocks.

Internal clocks profoundly affect behavior, and are found in virtually every living organism, whether bacterium, plant, or animal. In animals, mating activities, food seeking, sleeping, and waking are all tailored to the rhythms of the world in which they live. Seconds, minutes, hours, and months are all inventions of the human mind; the length of a day or a year are not. How these clocks operate—how they are set, and how they regulate behavior—has become one of the most active areas of current biological research. Where in the body are these clocks to be found? What cellular and molecular mechanisms underlie their function? Is there variability in clocks that explains variability in behavior, and if so, what role do genes play in this variability?

There are many kinds of biological clocks, and for the sake of convenience they are categorized by the lengths of time they measure. "Ultradian" clocks deal with periodicities in the range of seconds to minutes. Examples of ultradian clocks might include the pacemaker regulating normal heartbeat, or the oscillator defining wingbeat frequency in the mating songs of the fruit fly Drosophilia. "Circadian" clocks deal with activities attuned to the earth's normal twenty-four-hour cycle, such as sleeping. "Infradian" clocks regulate activities that occur with frequencies in the range of months to years; behaviors linked to the lunar cycle, or breeding patterns in certain large mammals, for example. These are useful categories for dividing up clocks, because the molecular mechanisms underlying each of these categories of timekeeping are likely to be different.

The clock with which we are most familiar is the one based on the standard terrestrial day of twenty-four hours: the circadian clock, from the Latin words for "approximately a day." This is an apt choice of words, because most circadian clocks actually run slightly longer or shorter than twenty-four hours. Humans, for example, have on average an inherent circadian period of just over twenty-four hours. Other organisms have cycles over, under, or essentially right at twenty-four hours. The difference between biological clock time and standard geological time is referred to as the "phase difference."

The phase difference in the human clock can be observed when humans are kept for long periods of time in continuous light or continuous dark. Under these conditions, our daily rhythms are controlled entirely by our internal clocks. We will still go to sleep and wake up approximately every twenty-four hours, and we will get hungry several times during our waking day. But in the absence of light changes, our clocks gradually slip out of synchrony with a true twenty-four-hour day. Soon we will be active while the rest of the world is asleep, and fall asleep while they are busy in the waking world. Under normal circumstances this slight phase shift is corrected by repeated daily exposure to the earth's light/dark rhythms, and when restored to a natural environment, light-deprived individuals quickly reset their clocks. But it is an intriguing and important feature of these clocks that, once set, they will continue to run, and to run the body's business, in the absence of the signal (light/dark changes) to which they are attuned.

Biological clocks are useful because they allow us to anticipate and prepare for scheduled activities before they actually begin. Our temperatures and metabolic rates are adjusted automatically throughout each circadian day, regardless of whether it is light or dark outside. Circadian clocks regulate a wide range of behaviors, often through hormones whose release is coordinated by the portion of the brain known as the thalamus. Sleep and wakefulness are the most obvious activities controlled by clocks, but clock-driven hormonal changes also affect major organ systems such as the kidneys and liver, as well as things like temperature, blood pressure, and heart and respiratory rates. The immune system and the body's ability to deal with chemical or electromagnetic insult are also regulated in part by circadian clocks. A great many of the perturbations in bodily functions brought about by clocks are expressed as overt changes in behavior.

Light-driven circadian clocks can be found in virtually every living organism exposed to light as part of its natural history, including even single-cell organisms. In fact one of the systems in which circadian rhythms have been most intensely examined is *Neurospora crassa*, a unicellular eukaryote familiar to us as the black mold growing on stale bread. Neurospora cells can readily be shown to possess the fundamental element of every clock: an oscillator. An oscillator allows clocks to divide time. In a pendulum clock, the pendulum itself is the oscillator that marks off time, depending on the length of the pendulum. In a spring-driven watch, energy provided by the wound spring is used to drive an escapement wheel, which acts as the oscillator. In Neurospora, as in all other living cells, the oscillator is provided by the protein products of specific genes.

We can see the oscillator at work in Neurospora cells by watching the way they produce spores, and by the effects of light on this process. If Neurospora cells are plated on agar and grown in the presence of continuous light, they grow out from the inoculation point in an ever-increasing circle. From time to time the Neurospora cells produce tiny spores or "conidia" which, unlike the rest of the growing cells, are pigmented and cause the black spots we see on moldy bread (Fig. 6.1A).

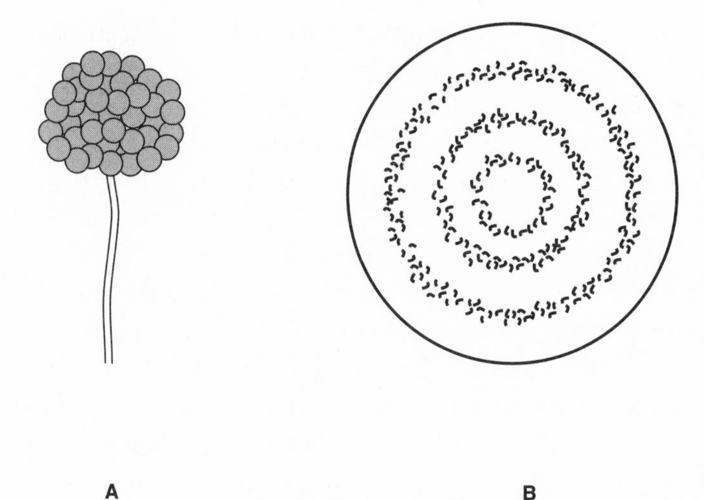

Figure 6.1 (A) An individual conidium showing spores clustered about a central stalk. (B) Neurospora growing outward from a central point on a Petri dish kept in the dark. The dish is uniformly covered with Neurospora cells, but conidia have formed only every 21.6 hours; the cells in between conidiation phases are colorless.

Conidia are produced more or less randomly through time, and so they are evenly spread over the surface of the growing circle. On the other hand, if Neurospora cells are grown in continuous dark, production of conidia suddenly becomes highly regulated through time. The cells grow outward from the inoculation point at a steady, unchanging rate, but conidia are produced only during a brief period of time; these "conidiation events" occur exactly every 21.6 hours. The result is a pattern of concentric rings of dark conidia superimposed on a continuous growth circle of Neurospora cells (Fig. 6.1B).

How do Neurospora cells keep time in the dark? How do they know when a new "day" has begun, and it is time to produce spores again? Why do we not see conidia rings when cells are grown in continuous light? As with many other biological phenomena in simpler organisms, insights into these questions first came from the identification of genetic mutations in Neurospora, specifically those that affect the periodicity of the conidiation cycle. The most important gene to be identified in connection with the Neurospora conidiation oscillator is called "frequency" (*frq*). So far, over half a dozen mutations of the *frq* gene have been described; some of these alleles of the *frq* gene increase the length of the circadian period (up to 29 hours), and some decrease it (down to 18 hours).

The *frq* gene encodes a protein (Frq)* that is produced in a cyclical fashion. At the Neurospora circadian equivalent of midnight, Frq is essentially absent from the cell, and the day begins by turning on its production. The concentration of Frq within the cell gradually increases, reaching a maximum level in early afternoon of the circadian period. The Frq protein moves into the nucleus, and as it reaches its maximum level, the Frq protein interacts with the *frq* gene and turns off its own production. The Frq protein then gradually disappears from the cell, and by "midnight" of the Neurospora circadian period, it has essentially disappeared. The absence of Frq from the cell allows the *frq* gene to begin operating again, and the cycle repeats.

The length of the circadian day is thus determined by the length of time required to synthesize enough Frq protein to turn off the *frq* gene, and by the length of time required to clear existing Frq from the cell and allow reexpression of the *frq* gene. This is the molecular essence of the Neurospora circadian oscillator. Light definitely affects this process; Neurospora cells grown in continuous light produce conidia on a continuous basis rather than cyclically. Direct measurements have shown that light exposure of ninety seconds or more stimulates the production of Frq protein, and the increase in Frq is directly proportional to the amount of light in the pulse. This provides an obvious basis for explaining the effect of exposure to light on conidiation, if we assume that the Frq protein stimulates conidiation, and that continuous light induction of Frq overcomes Frq self-suppression. It is clear that in the presence of continuous light, Frq is kept at a constantly high level in the cell, and this continuously drives conidiation.

The effect of light on Frq levels in the cell could also explain another process characteristic of virtually all circadian clocks: the ability of light to reset them. If Neurospora cells growing in the dark are exposed to a pulse of light in the early part of their circadian day, the concentration of Frq begins to increase in the cell. The increased concentration of Frq causes the cell to reach "noontime" earlier than normal, thus

---

* When the nature of the protein encoded by a defined gene is not known, it is given the same name as the gene, but to distinguish between the two, the name of the protein is written in roman letters, with the first letter capitalized. The name of the gene is given in lowercase italic letters.

shortening the day (Fig. 6.2). If the light pulse is given in the circadian afternoon, when Frq is decreasing in the cell, there is a temporary increase in Frq, thus slowing the afternoon and increasing the circadian day. Once the clock has been reset by a light pulse, the circadian rhythm carries forward with the same cycle length as before, but slightly shifted in terms of external "earth" time. Resetting the clock is important for animals and plants because the lengths of day and night vary throughout the year, and clocks must be continually readjusted. This process is also related to the effect of light on humans recovering from jet lag, as we discuss below.

Now that the oscillator is well on its way to being defined in Neurospora, scientists are working on two important questions related to clock function. How does light shining on the surface of a cell affect processes going on inside the cell? And how does the clock, once set in motion by the oscillator, act to control events inside the cell?

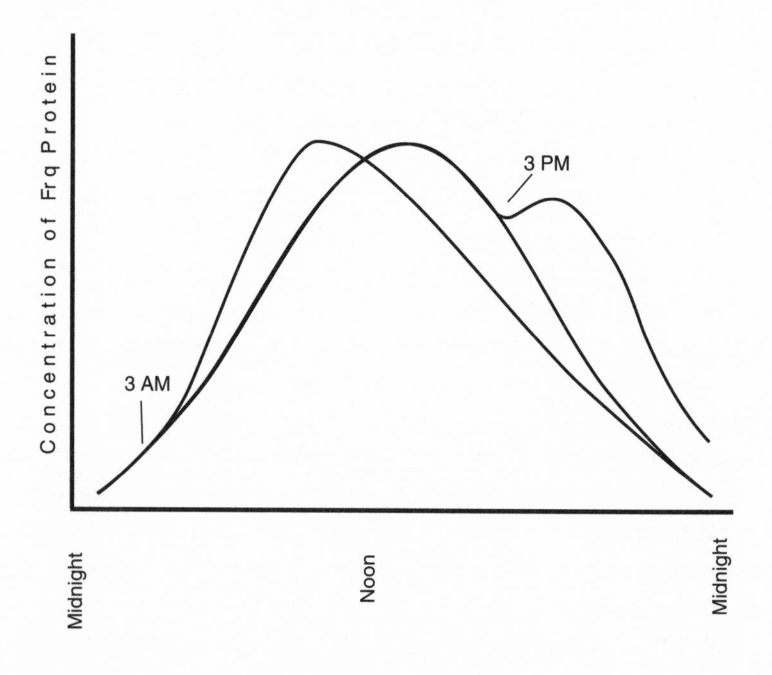

Figure 6.2 The concentration of the Frq protein within Neurospora cells grown in the dark, as a function of time during the circadian day. The normal distribution pattern (middle curve) can be shifted earlier or later by exposure to light at 3 a.m. or 3 p.m., respectively.

Light is detected in cells by a special category of proteins called photoreceptors. Their job is to transmit light signals to the clock oscillator. In multicellular animals, photoreceptors are partitioned off into specialized cells serving as eyes. In Neurospora, as in other unicellular organisms, all of the functions of a multicellular organism—respiration, nerve function, interaction with environmental light and other signals—are combined in a single cell. Some photoreceptor proteins are embedded in the membrane of the cells, but in Neurospora, as in plants, the photoreceptor protein is located in the cytoplasm. Photoreceptor proteins have special chemical groups called chromophores attached to them that undergo a chemical change when struck by a photon of light. In standard mammalian vision, this chromophore is a form of vitamin A, which is why this vitamin is important for vision. The chromophore in Neurospora is likely to be a form of vitamin B. Activation by light through the chromophore usually involves a slight change in the configuration of the photoreceptor protein, subtly altering its activity.

What happens once the clock is set in motion is also an intense area of study. Waxing and waning of Frq is now established as the oscillator driving the Neurospora clock, but how does the clock connect to other biological processes going on in the cell? The Frq protein is presumed to regulate genes involved in conidiation by interacting with these genes to turn them on or off. The evidence that Frq can interact with the *frq* gene itself and turn it off is quite strong, and it is entirely possible that Frq itself interacts, either directly or indirectly, with other genes as well. We know that continuous light, which induces a constant high level of Frq, results in continuous conidiation, so we can guess that Frq may induce expression of the relevant genes. Studies of the structure of Frq, inferred from the sequence of the *frq* gene, suggest it may be among those proteins that interact with DNA to regulate gene expression. A series of genes called *ccg* (clock-controlled genes) have been identified that are selectively controlled—either activated or repressed in a time-dependent fashion—by the Neurospora clock. Eight have been identified so far, and there will almost certainly be many more. We do not yet know what these genes encode, and what their protein products do within the cell, but this will likely be worked out in the next few years.

So clocks were already present in single-cell organisms, probably a billion years or so before multicellular organisms ever arose on the earth. What happened when life became multicellular? Can all of the cells of a multicellular organism tell time, or is this function, like so many others, delegated to only certain subsets of cells? Virtually all of the elements of biological clocks found in the most complex animals were already present in Neurospora, and as we see below, the genes responsible for these elements in humans are not terribly different from the genes in a simple bread mold. But where are these genes expressed? To see what happened to time-keeping duties in the course of evolution, let's take a look at an intermediate level of organization of clocks, in an organism in which clocks have been dissected in as much detail as Neurospora: the fruit fly *Drosophila melanogaster.*

Keeping Time in Flies

The potential importance of biological clocks in regulating animal activities led Seymour Benzer to explore the genetic basis of clock function early in his studies of behavior in Drosophila. As multicellular animals with a distinct nervous system, flies are involved in a wider range of activities than Neurospora, and the involvement of clocks in Drosophila is correspondingly broader and more complex. When kept on a standard twelve hours light, twelve hours dark cycle, two behaviors of the fly are easily seen not to be randomly distributed in time. In a process called "eclosion," fully formed flies emerge from their pupal cases preferentially just after the light comes on ("dawn"), and this activity wanes throughout the rest of the day, essentially ending by the time the lights are turned off ("sundown"). Locomotor activity—the tendency of flies to fly or walk about—occurs most actively just after dawn and just before sundown. But for both activities, there is a period of anticipation of changes in light: Just before dawn, some flies begin to emerge from the pupae and some flies begin to move about. Some flies also move about just after sundown, but then this activity essentially stops.

That this behavior is regulated by an internal clock is suggested by several observations. For example, the circadian patterns of both behaviors can be reset by changing the light cycle. However much the light cycle is shifted, both behaviors still follow the same dawn/sundown pattern. This could be attributed simply to a direct effect of sunlight on the behaviors involved, but flies maintain this rhythm in complete darkness. The fly circadian pattern exhibited in the dark is actually very close to twenty-four hours, with dawn and sundown behaviors occurring on the same schedule as if light were present. But the final proof that these behaviors were governed by an internal clock came with the discovery by Benzer's student Ron Konopka that both are affected by a mutation in a single gene, called *period*. Allelic variants of the period (*per*) gene alter the rhythmicity of both behaviors by changing the circadian day length. In their initial studies, Konopka and Benzer found three allelic mutations of *per*; one that lengthened the day (per^l) to about 28.5 hours, one that shortened the day (per^s) to about 19.5 hours, and one (per^0) that resulted in a complete loss of rhythmicity. In the dark, the per^0 mutants walk and fly around randomly throughout the circadian day and night, and per^0 flies emerge from pupal cases without respect to circadian time. This was the very first of what would eventually turn out to be hundreds of behavioral mutants in Drosophila.

In their classic 1971 paper, Konopka and Benzer proposed that the *per* gene controls an important component of the Drosophila oscillator. Although the level of the Per protein was later shown to cycle up and down during the circadian day, there was some difficulty in correlating circadian rhythmicity directly with activity of the *per* gene by itself. These difficulties were resolved with the subsequent identification of a separate mutation called "timeless," the underlying gene for which is called *tim*. Timeless mutants, like per^0 mutants, show a complete loss of circadian rhythm, but the *tim* gene is separate and distinct from *per*.

The important feature of the *timeless* mutation is that it also blocks the circadian cycling of Per within the cell. This led to the conclusion that *per* and *tim* work together to control the Drosophila circadian day (Fig. 6.3), through a required interaction of the Per and Tim proteins. In a normal twelve-hour light, twelve-hour dark cycle, these proteins are synthesized, like all other proteins, in the cytoplasm. During periods of light, the levels of Tim are kept low, but toward dusk, the levels of Tim begin to build up. Tim then begins to pair up with Per, and the dimers enter the nucleus where, like Frq in Neurospora, they turn off the genes that encode them. As a result, levels of Tim and Per in the cytoplasm start to fall, and Tim is further destroyed by light. When

Tim–Per pairs also begin to disappear from the nucleus, synthesis of Tim and Per start up again. In a *tim* mutant, where there is no Tim protein (Fig. 6.3B), Per is synthesized, but remains in the cytoplasm, unable to enter the nucleus because it cannot pair with Tim. In a *per* mutant (not shown), Tim is synthesized, but would be continually destroyed by light.

The basic mechanisms of the oscillators driving the Drosophila and Neurospora clocks are thus seen to be very similar at the molecular level. In both cases, time is measured by the cyclic generation and destruction of proteins encoded by specific genes dedicated to that function. In flies, as in bread mold, the clock operates entirely at the

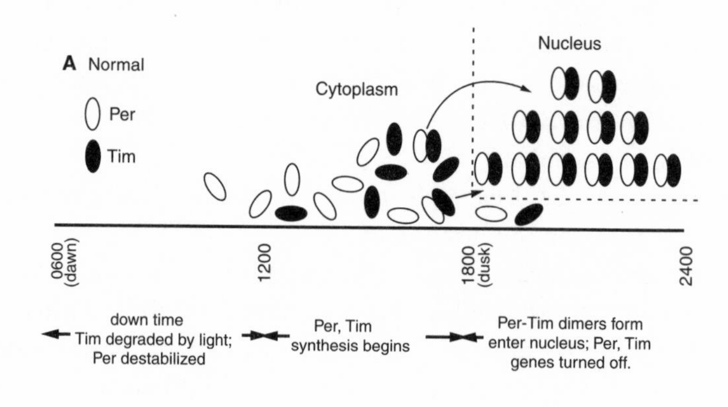

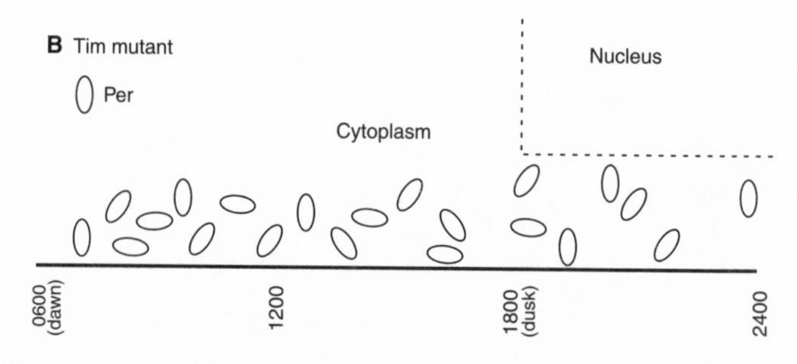

Figure 6.3 (A) Normal cycling pattern of Per and Tim proteins in Drosophila cells exposed to a twelve-hour light/dark cycle. (B) Per proteins in *tim*-mutant cells.

cellular level. In Neurospora, the cell is the organism, so location of the clock within the organism is not a question. But Drosophila are highly organized multicellular animals, with different functions compartmentalized into different cells and tissues. One important question we should ask about Drosophila is, where in the flies is the clock located? Individual cells have clocks, but in multicellular organisms is the clock unicellular or multicellular? If multiple cells are involved, do they function independently or cooperatively?

Since all circadian clocks are affected by light, one obvious location for clocks would be somewhere in the nervous system, perhaps in that portion of the brain that receives and processes light signals. And indeed, the Per protein can be found both in Drosophila eye structures and in brain tissue, where it can be seen to cycle on a daily basis. This initially led investigators to suppose that a central clock in Drosophila brain controlled cyclical behaviors throughout the entire fly, perhaps by altering the levels of brain-controlled hormones. But that rapidly turned out not to be the case; when these investigators looked further in flies, they found that Per is actually present and cycling in almost every cell type they examined!

Although the brain-centered clock undoubtedly does have a major influence on regulating overall circadian behaviors, it now seems that most fly tissues have their own internal clocks to assist the central clock in regulating local activities. That these clocks are truly independent was demonstrated in experiments where different parts of a fly's body were cut out and placed in separate culture dishes under conditions where they could be maintained in a living state for several days.* It was shown clearly that circadian cycling of oscillator proteins could be maintained in the dark in all of these tissues, in the absence of connection with a brain, and that their cycles could even be independently reset to a new rhythm through selected exposure to light. More recently, the existence of clocks in individual cells has been shown for

---

* This is not as strange as it may seem. Flies are very small, and segments of a fly body are no larger than tiny pieces of human tissue that can also be removed from the body and maintained alive in culture for several days or more. The mammalian heart, for example, beats more rapidly during the day than at night. If heart cells are dissociated into individual cells and placed into culture, they will continue to beat according to this same circadian rhythm for as long as they can be kept alive.

all of these tissues, regulating a wide range of different cyclical activities throughout the body.

Clearly the eyes are not funneling a light signal to separated body parts grown in culture dishes. So how do individual cells pick up a light signal? The same way that Neurospora cells do: through the mediation of intracellular photoreceptors. Although clearly multicellular, Drosophila are still rather tiny, and light can penetrate most of the way into the Drosophila body. In flies, as in people, only retinal cells have the classic light-gathering proteins associated with vision, but the presence of blue-light-sensitive photoreceptors has now been detected in a wide range of non-neural cells. Although these receptors have not yet been linked formally to clock function in Drosophila, it seems very likely that they will be. The notion that clock function may be spread throughout the entire body, rather than encased in a single, central pacemaker, is changing the entire framework within which scientists think about circadian rhythms.

Circadian Rhythms in Mammals

Mammalian clocks govern a number of bodily activities directly affecting behavior, for example, daily body temperature shifts, sleep/wake cycles, mental alertness, and physical activity. But even the smallest mammal is thousands of times larger than an adult Drosophila, and the vast majority of cells are buried deep below the surface, inaccessible to environmental light. Thus, while circadian rhythms are abundantly evident in mammals, they are much more dependent on the light-gathering capability of the eye for their regulation. In mammals, including humans, there is a region of the brain called the suprachiasmatic nucleus (SCN) that has long been known to play a major role in establishing and maintaining circadian rhythms. Located within the hormone-producing hypothalamus, the SCN has direct neuronal connections to the light-gathering retinal cells in the eye, and removal of the SCN in rodents effectively destroys rhythmic behavior.

The importance of the SCN in regulating mammalian circadian behaviors was shown rather dramatically in experiments with hamsters. Hamsters normally operate on a circadian rhythm very close to twenty-four hours, and this cycle regulates their exercise patterns in a running wheel cage; they tend to exercise at the same time each day, even when kept in the dark. However, a mutant substrain of hamsters has been established in which the circadian period is twenty hours; when kept in either constant light or constant dark, they quickly slip out of phase with normal hamsters kept under the same conditions, as expressed by their internally regulated exercise schedule.

Using a very carefully inserted electrical heating element, researchers destroyed the SCN in the mutant strain. The animals were allowed to recover, and when subsequently tested showed that they had completely lost their circadian clock. The researchers then implanted SCN tissue from embryos of a normal hamster into a region of the brain very close to the destroyed SCN. The transplanted mutant hamster now displayed a normal twenty-four-hour circadian cycle. Cortical brain tissue implanted in the same situation failed to restore any clock function at all. When the reverse experiment was performed, normal mice transplanted with a mutant SCN displayed a twenty-hour circadian rhythm. This experiment shows beyond doubt that the SCN in mammals is a major regulator of daily behavioral activities.

Exactly how the light used to regulate mammalian circadian rhythms is gathered was something of a puzzle until recently. If the nerve fibers leading from both eyes to the brain are cut, standard vision is of course completely lost. The internal clock, having been previously set by ambient light, continues to run. But it can no longer be reset by external light, and so, if the internal clock is not tuned to exactly twenty-four hours, the individual gradually slips out of synchrony with the standard day, as would happen with a fully sighted individual kept in constant darkness. The important point is that the clock in such subjects can no longer be reset by exposure to light, which tells us that the eye is absolutely critical in maintaining the clock.

The cells within the retina of the eye that gather light for vision are the rods and cones. Rod cells are dispersed evenly across the retina, and use a photopigment called rhodopsin to gather light. Rod cells are used for vision in dim light, and discriminate shades of gray rather than color. Color vision is the task of the less numerous and centrally clustered cone cells, which use three different pigments, based on three different variants of a common opsin protein, to detect long-, medium-, and short-wave light (basically, red, yellow, and blue). The light-trapping ligand for all of the opsins is cis-retinal, a form of vitamin A. The genes for the various color-detecting opsin proteins are all variants of a single opsin gene that arose in animals about 800 million years ago.

It is clear that the light used for daily resetting of mammalian clocks controlled by the SCN enters through the eye. Yet in a strain of mice in which the rods and the cones extensively degenerate, with essentially no ability to gather light, circadian rhythms not only are maintained, but can be as easily reset by light as in fully sighted mice. Moreover, most blind humans lose the ability to reset their clocks to the daily light cycle, and operate entirely on their internal circadian rhythm. This can lead to recurrent problems of falling asleep during the day while being unable to sleep at night. Normal sleep patterns are gradually restored as their circadian day realigns with the external light/dark cycle, but then they gradually fall out of phase again, over and over. These individuals also do not experience jet lag when they cross multiple time zones in rapid air travel. Yet other blind people, with absolutely no ability to gather light for purposes of seeing, have perfectly functional circadian clocks that keep them on a true twenty-four-hour schedule, synchronizing with daily light cycles and seasonal changes in light/dark patterns. These individuals suffer jet lag as much as fully sighted people.

So how do mice and some humans who are blind maintain circadian rhythms that can be reset by external light? The resolution of this dilemma came about recently when it was shown that light used to drive mammalian circadian clocks is not harvested by the rods and cones that gather light used for standard vision. Researchers have discovered a completely separate light-gathering system deep within the mammalian eye, dedicated exclusively to regulation of the circadian clock. It is a blue-light photoreceptor pigment called cryptochrome, and consists of a protein carrier and a form of vitamin B2 as the light-trapping ligand. Cryptochrome is the chromophore used in Drosophila, and may well be related to the blue-light-gathering photoreceptor system in Neurospora. It is closely related to the system in plants that gathers light for managing plant circadian rhythms. Thus, just as with mammalian olfaction, where different sensory systems pick up volatile chemicals serving as signals for smell (olfactory neurons), and as with behavior-altering pheromones that stimulate vomeronasal neurons, there appear to be separate systems for detecting light used for standard vision, and for behavior-altering circadian clocks.

The discovery of what is essentially a plant light-gathering system operating in mammals (including humans) was a major surprise. Plants and animals were separated in evolution at least a billion years ago, and it seemed as though genes for structures like cryptochrome had simply been left behind as animal life emerged. Plants have two genes for cryptochromes, called *cry-1* and *cry-2*. Both of these genes were found to be present in humans and mice. Using the Drosophila genes as probes, researchers found that both of these genes, like *per* in flies, are not only present but actively expressed in most cells of the body in mammals, including humans. Most important, they are also expressed in both the eye and the SCN region of the brain. These genes turn out to be expressed in different cells within the retina than are the genes for the standard visual pigments, which helps explain how some individuals who have lost rod and cone cell function can still reset their clocks with external light. In the SCN, expression of the *cry-1* gene was found to oscillate in an obvious circadian pattern, with optimal expression during periods of light, and undetectable expression in the dark.

But the SCN-based circadian clock may not be the whole story. In a recent paper published in the journal *Science*, researchers described resetting of human circadian rhythms by light directed not at the retina, but behind the knee! Two clock-regulated parameters were monitored in these studies: body temperature and blood levels of the hormone melatonin. Human body temperature drops by about half a degree Celsius each night, reaching a minimum between 2 and 6 a.m., after which it returns to normal by about noon. This pattern persists in the absence of light/dark changes, in which case it synchronizes to the internal circadian clock rather than ambient light. Circulating melatonin levels have a reciprocal pattern. Melatonin is a hormone produced in the pineal gland,* which is connected through nerve fibers to the SCN. Peak melatonin production normally occurs in the middle of the night, at just about the same time body temperature reaches a minimum. External melatonin administered in the daytime can cause

* The pineal gland is not actually part of the brain per se. In some lower vertebrates, it has structural similarities to an eye, and may actually play some role in gathering light. In mammals, it resembles more an endocrine gland, and has no known light-gathering functions, embedded as it is deep within the braincase.

a drop of about half a degree in body temperature, so it is entirely possible that elevated melatonin causes, rather than simply correlates with, lower night-time body temperature.

Both body temperature changes and melatonin patterns can be reset by ambient, whole-body light administered in normally dark periods, in effect resetting the circadian clock. In the study published in *Science*, it was found that these two circadian parameters can also be reset by very high intensity light administered behind the knee. Healthy subjects were placed on a strict light/dark schedule, and their baseline patterns of body temperature and circulating melatonin levels determined. They were then exposed to bright light for three hours, using a covered fiber optic pad placed behind the knee, with the lower part of their body shielded to prevent any possibility of light leakage to the eyes. By administering the light pulse at different parts of the circadian cycle, the researchers were able either to advance or retard the circadian cycle (Fig. 6.4).

How light administered to the body surface rather than through the eyes could affect a clock located in the brain is unclear at present. It has been suggested that bright light directed onto skin may cause

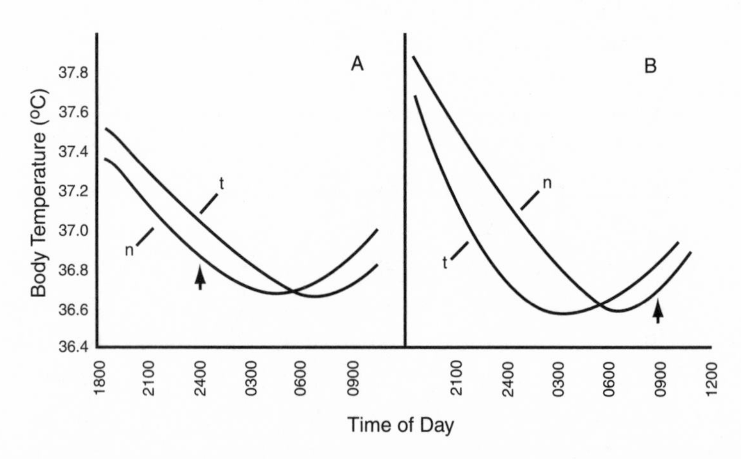

Figure 6.4 Circadian phase shifts in humans caused by light applied behind the knee. (A) The normal (n) nightly minimum in body temperature is shifted in time about three hours in a subject treated (t) with light prior to the minimum temperature. (B) The normal (n) nightly minimum is delayed in a subject treated (t) with light after reaching the minimum temperature. Based on data presented in S. Campbell and P. Murphy, *Science* 279:396–399.

changes in blood gases passing through capillaries just below the surface of the skin, which could travel to and affect a clock located in the brain. But we also know now that the blue-light photopigment is distributed in cells throughout the body. It is possible that a chemical messenger is produced within cells located at the body surface that have been subjected to intense light. This messenger could be released into the bloodstream and make its way to the SCN.

Although other body sites have not been tested, there is no reason to believe that the back side of the knee is a uniquely photosensitive site. Most studies have suggested that normal daylight, and levels of artificial light in homes and workplaces, has no discernible effect on human circadian rhythms unless it impinges on the retina. For normal daily activities, the eye is essential for gathering the light that controls clock function. Yet it remains intriguing that the photoreceptor used by the eye to gather light for our biological clocks is spread throughout the body. Whether this photoreceptor plays a subtle role in detecting seasonal changes, or perhaps a role completely unrelated to light gathering, remains to be seen.

In addition to the *cry-1* and *-2* genes, mammals have also retained the *per* gene described in Drosophila. In 1997, two laboratories reported the isolation and cloning of *per* genes from mice and humans. The mouse and human genes are 92 percent alike, and both are about 45 percent identical to the Drosophila per gene. The mouse gene is found in the SCN, where its expression oscillates in a circadian pattern that persists in the dark and that can be reset by light pulses. Another mouse gene called "clock" has recently been cloned; its protein product appears to facilitate the interaction of Per protein with DNA. All indications at present are that the basic molecular machinery present in clocks typical of the earliest eukaryotes are conserved in present-day mammals.

Clocks, Genes, and Human Behavior

It seems almost certain that the mechanisms governing clocks in organisms like Neurospora and Drosophila are used in human clocks as well. The genes and proteins driving circadian clocks are the same in virtually every organism that has been examined, and these same genes and proteins are present in humans. It seems highly unlikely that humans have retained them but use some other, completely unknown mechanism to operate their clocks. Although it remains to be proved formally, we imagine that the human clock is under control of an oscillator that consists of a cyclically generated and destroyed protein, quite likely the product of the recently identified human *per* gene. Other proteins will almost certainly be involved, and all evidence at present indicates that these proteins, in addition to regulating their own expression and destruction, are involved in turning on and off the so-called clock-controlled genes that are directly responsible for rhythmically cycling behaviors.

One of the best indications that clocks affect human behavior comes from the study of psychological disorders. Persons suffering from chronic depression display numerous behavioral alterations, particularly in their interactions with others, and they also show perturbations of their normal clock functions. They nearly always experience their most severe bouts of depression in the morning, usually after disruption of sleep just before dawn. Melatonin production is unusually low in many depressed patients, and some antidepressant drugs not only restore melatonin production, but also advance the circadian clock somewhat. Lithium, used to treat manic-depressive disorder, also advances the circadian clock. One of the clearest involvements of clocks in depression is in seasonal affective disorder, where individuals have moderate to severe bouts of depression only during winter. Phototherapy—brief exposure to bright light at specific times in the circadian cycle—has been used with some success to stabilize and reset circadian rhythms in these individuals.

As with all biological clocks, human clocks can be reset by ambient light. The most obvious example of this is jet lag. Anyone who has crossed more than three or four time zones on a fast-flying airplane is familiar with jet lag—one of those biological surprises revealed only through the development of modern technology. Jet-lagged individuals are fatigued, dehydrated, and stressed, but most important, their clocks are markedly out of sync with the light/dark pattern in their new location. They may recover from the stress and fatigue of flying within a matter of hours. Resetting internal clocks, and establishing new eating and sleeping schedules, can take days. Similar difficulties are experienced by nightshift workers who are rarely able to reset their clocks because they also experience various exposures to light during their sleeping periods, and then flip clocks completely on weekends. Astronauts may experience particularly difficult internal clock problems.

In studying jet lag more closely, it has been found that human clocks reset more readily when we fly westward. For flights across five or more time zones, our clocks reset at a rate of about ninety minutes a day for westward flights, but at a rate of only about sixty minutes a day on eastbound flights. These adjustment rates apply whether the flight was outbound or homebound, whether flying in the daytime or at night. The reason for different clock recovery rates in eastbound versus westbound flights is readily understandable in terms of human circadian rhythm. Most humans have a built-in circadian rhythm slightly longer than twenty-four hours. Traveling eastward, clocks readjust by phase advances, that is, by shortening the circadian day. In westbound travel, clocks must temporarily lengthen the circadian day, until they catch up with local time. Since the circadian day is already slightly longer than twenty-four hours in humans, our clocks have a slight advantage in catching up when we travel westward.

Considerable attention has been paid recently to the use of both light and melatonin for treating jet lag. The rationale for phototherapy is clear, given the extensive documentation in animals and in humans that light can reset circadian rhythms. However, the use of melatonin as a medicament has been controversial. Melatonin is of course a marker of the circadian cycle, and its synthesis in the pineal gland is depressed by light, presumably secondary to a signal received through the SCN. But it turns out that cells in the SCN themselves have receptors for melatonin, and so communication between the pineal gland and the SCN may occur in both directions. In fact, melatonin has actually been used to reset circadian rhythms in animals and humans, and has been used in the treatment of various depression syndromes, including seasonal affective disorder. Although not acutely toxic, even in high doses, it must be used with great caution, since the implications of long-term use of supplemental melatonin are not completely understood at present.

As mentioned earlier, many blind people have also lost their ability to gather light via the blue-light photoreceptors dedicated to clock function. A recent detailed study of a blind patient in Germany showed that his circadian day was twenty-four hours and fifteen minutes long. As a result, every three to four months, he would be twelve hours out of phase with the standard earth day, and would have difficulty staying awake during the day. At the same time, he was unable to sleep at night. This was a highly intelligent individual who understood fully his problem. He kept meticulous records of his circadian rhythms, and worked hard to adjust his daily activities based on nonvisual timing cues from his environment—traffic sounds, radio programs, the activities of others around him. But he was never able to do so. His experiences are similar to those of thousands of other blind persons, who have used many different techniques including hypnosis and meditation in futile attempts to reset their clocks. This tells us that the human circadian clock cannot be adjusted by psychological means, but only by light or melatonin.

On the other hand, the recent findings with respect to the distribution of the blue-light photoreceptor throughout the body, and the demonstration that the human clock can be reset by intense light trained on body sites other than the eye, offers some rather interesting possibilities for human biology. The potential for intense localized light exposure, as well as generalized whole body phototherapy, to reset clocks shifted by jet lag is already being explored. The even more valuable potential for such light treatments in helping time-shifted blind persons to reset their clocks to normal daily rhythms will surely be explored in the very near future.

Circadian clocks are one of the most fundamental regulators of behavior in all of biology. Every aspect of the behavior of animals—including humans—is cued by where they perceive they are in the course of a normal day. Even bacteria and plants have clocks, and these clocks work in exactly the same way that animal clocks do, although the genes and proteins involved are completely different. This suggests that clock systems may have arisen multiple times in evolution, completely independently of one another, but each time arriving at the same basic oscillator mechanism. Given the pace of research in this field, we will likely have most of the components of mammalian clocks identified in the next few years. We can then ask to what extent the genes encoding these components vary within the human population. The genes for most of the clock components described in other mammals so far occur in multiple forms, and it is likely that the corresponding human genes will as well. We will then want to know how variations in these different clock components affect behaviors controlled by circadian clocks. The involvement of circadian rhythms in certain psychological disorders suggests that this could well be one of the more fruitful fields of analysis in understanding the genetic basis of human behavior.

7

You Must Remember This

The Evolution of Learning and Memory

In humans, as in all animals, behavior is orchestrated almost entirely by the central nervous system: the brain, where it exists, and the scattered collections of neurons known as ganglia. Of all the functions of the central nervous system, by far the most crucial for behavior is the ability to learn and to remember. Learning is the impression on the nervous system of information about both the inner and the outer worlds. Memory is the storage of that information for future use. It is what allows us to compare a given experience with what has gone before, and to respond more effectively to future situations. It is the means by which the environment exerts its impact on behavior.

Intuitively, scientists have always felt that memory must be based on some sort of change in the brain itself, as a result of experience, but the exact nature of that change has only recently begun to reveal itself. We now know that memory results from both physical and chemical alterations in those nerve cells that have been affected by an environmental cue. The acquisition of memory involves two kinds of changes in the brain. One change lies in the molecular pathways for information processing that exist inside each neuron. The other change involves the way nerve cells interact with each other; both the nature and number of synapses can be altered by environmental input. As is so often the case, our most important insights into the nature of these changes have come not from studying the human brain itself, but from watching the changes unfold in much simpler organisms—in this case, a distant relative of a common garden pest.

Learning and Memory in a Sea Slug

Aplysia californica is a marine slug, a snail relative that spends its entire life in the sea, eating, reproducing, and absorbing oxygen from seawater through a gill that protrudes from its back (Fig. 7.1). Although more complex than *C. elegans*, Aplysia has a number of advantages as a model system. Its nervous system is readily accessible, and the individual nerve cells are rather large and easy to work with. The nervous system contains about 20,000 cells, compared with the 302 nerve cells in an adult *C. elegans*—too many to trace the fate of every single neuron, yet simple enough so that a number of key neuronal interactions have been identified. The neurons are organized into distinct ganglia, somewhat akin to the neck ring in *C. elegans*—basically, a place where neurons meet and confer. In Aplysia, a number of these ganglia have taken on at least some of the characteristics of a primitive brain.

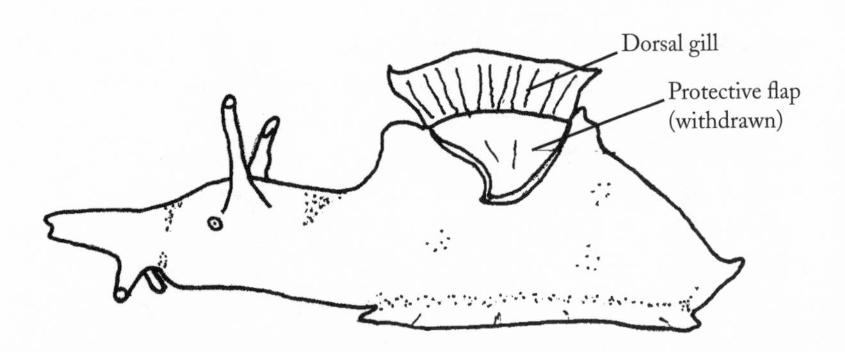

Figure 7.1 *Aplysia californica*, a marine sea slug.

Unlike *C. elegans*, which was purposely selected for ease of genetic manipulation, the value of Aplysia has not been the identification of mutations that affect nervous system function. It is not used to study heritability or even behavior per se. Rather, Aplysia has been used to work out the detailed cellular and biochemical pathways involved in learning and memory. There are two fundamental types of memory: implicit (nondeclarative) memory, and explicit (declarative) memory. Each of these in turn has a short-term and a long-term component. Implicit memory is the most basic type of memory, and it mostly involves routine, essentially reflexive neuromuscular responses to environmental stimuli. It is definitely a form of memory, but there is no "thinking" involved, and no images associated with previous experience are summoned forth. Declarative memory, on the other hand, involves the recall of abstract mental images about individuals or events, or learned information. It is what we humans have in mind when we speak of memory.

Aplysia, like *C. elegans*, is capable only of implicit memory; we don't really begin to see explicit or declarative memory until we reach the vertebrate stage of evolution of the brain and nervous system. Nevertheless, vertebrates (including humans) also have implicit memory, and there is every reason to believe that the basic mechanics of implicit memory are the same throughout evolution, from *C. elegans* to humans. The major goal of the Aplysia work is nothing less than the identification of every molecular pathway involved in the perception of a signal by the nervous system, its integration with previous experience, and its conversion into both short- and long-term implicit memory. Ultimately this means identifying every single protein in these pathways, and figuring out just what each protein is doing.

So far all of the proteins identified as being involved in implicit memory have been well known to biochemists for some years. They are proteins involved in workaday tasks in other cells of the body; few are brain-specific. The unique value of the Aplysia studies comes from gaining an understanding of how all of these proteins work together to generate, store, and retrieve memories, and how memory can be used to modify subsequent behavior. For behavioral geneticists interested in the variability of inherited behavioral traits, the proteins identified in the Aplysia work are a rich resource. It will be possible to work backward from the proteins identified in Aplysia to the underlying genes, many of which in fact have already been identified. One can then ask whether variability in these genes is inherited along with variability in a given trait. Conversely, if one identifies a gene involved in a behavior, and it turns out to encode one of the proteins identified in Aplysia, we will know exactly what it does and how it works.

Because nature is so conservative at the level of the genes and molecules used to operate cells, the information gleaned in Aplysia will in the vast majority of cases apply to other organisms, including humans. So let's take just a brief look at the major findings in Aplysia; we see many of the very same features involved in learning and memory in higher organisms as we proceed through this book.

Like *C. elegans*, Aplysia responds to its environment through various types of sensory neurons distributed over the surface of its body. For example, the dorsal gill and associated structures are equipped with a couple of dozen or so surface nerves that are highly sensitive to pressure (Fig. 7.1). If a strong jet of water is directed against the gill, it is immediately withdrawn into the body. This is a classic example of a reflexive behavior. The sensory nerves pass a signal directly to motor nerves innervating a tiny muscle just below the gill. The animal senses something unusual—and therefore potentially threatening—in the environment, and takes appropriate action, in this case withdrawing its gill.

As in *C. elegans*, this response can be habituated by experience. But in Aplysia, we can actually see the molecular events that accompany this process. Repeated activation of the sensory neurons leads to gradual inactivation of a key calcium ion channel within those neurons, which in turn results in less neurotransmitter released by the sensory neuron to the motor neurons to which it is synaptically connected. At some point not enough neurotransmitter (a molecule called glutamate, in this case) is released to drive the withdrawal response. Habituation in this instance results from shutting down a key neuronal function.

A simple habituation state is usually induced by ten or so consecutive disturbances of the gill, spaced a few seconds apart. Recovery from such a habituated state requires only a few minutes, and so habituation can be thought of as a form of short-term memory. But if a slug is exposed to several sequential sets of these repeated stimuli, a state of habituation can be induced that lasts for several weeks, and may represent a form of long-term memory. In this state, not only is neurotransmitter release from the sensory neuron decreased; some of the synaptic connections made by the sensory neuron are actually disrupted. This "remodeling" of synaptic connectivity is an example of the second change taking place during memory acquisition. It is now thought to be a common physical feature in the development of both implicit and explicit long-term memory.

Aplysia also exhibit a more advanced form of learning called "sensitization." When organisms are exposed to a potentially harmful stimulus, large portions of their sensory neurons go into a heightened state of alert, and can remain there for some time. Having learned that there may be danger in their immediate environment, the entire sensory system is alerted to possible threats from without. For example, if Aplysia are given several electrical shocks at a point far removed from the gill, and then later the gill is given a tactile stimulus, the withdrawal reflex is much stronger than in animals not receiving prior shocks. But the sensory neurons detecting the electrical shock, and the sensory neurons detecting tactile stimulation of the gill, are completely separate. How can stimulation of one set of nerve endings affect the response of totally separate and remote sensory neurons?

The answer lies in a special type of interneuron called a facilitating neuron, which marks a major step forward in the evolution of nervous systems. The electrical shock results in the recruitment of facilitating neurons, which now send out fibers to many other sensory neurons in the area, regardless of what it is they sense: light, pain, heat, and so on. As a result, these sensory neurons are all placed in a heightened state of alert, and will respond more strongly the next time they are stimulated by whatever it is they detect. Moreover, whereas during habituation the sensory neurons serving the gill disconnect somewhat from the motor neurons they serve, the facilitating neurons actually stimulate an increase in the number of synaptic connections made by sensory neurons in their various pathways. This, too, is part of the establishment of long-term memory, and again we see a physical remodeling of synaptic connections. But the major shift in sophistication in these kinds of responses is that neurons that previously were not in connection with one another are now bridged by facilitating interneurons, and messages are flowing back and forth through new pathways.

Facilitating neurons release a number of neurotransmitters, among which is one called serotonin. Serotonin is picked up by other neurons through a cell-surface receptor. As a result of binding serotonin, the sensitivity of sensory neurons to stimulation is greatly increased. Serotonin receptors on sensory neurons, when occupied by serotonin, indirectly trigger the conversion of a molecule called ATP (one of the building blocks of DNA, here playing a completely different role) into an intracellular molecule called cyclic AMP (cAMP). This latter molecule acts as a so-called second messenger. It helps relay information from a cell-surface receptor to the interior of the cell (Fig. 7.2). Cyclic AMP is by no means restricted to nerve cells; it is used in many cells throughout the body in the mediation of many types of signals.

One of the roles of cAMP is to bind to and activate a protein called a "kinase." In the establishment of short-term memory, the activated kinase induces changes in cell-surface potassium ion channels (just like those in paramecia) that lengthen and amplify incoming signals. The kinase also promotes a more rapid release of neurotransmitter-containing granules from the sensory neuron. In the transition to long-

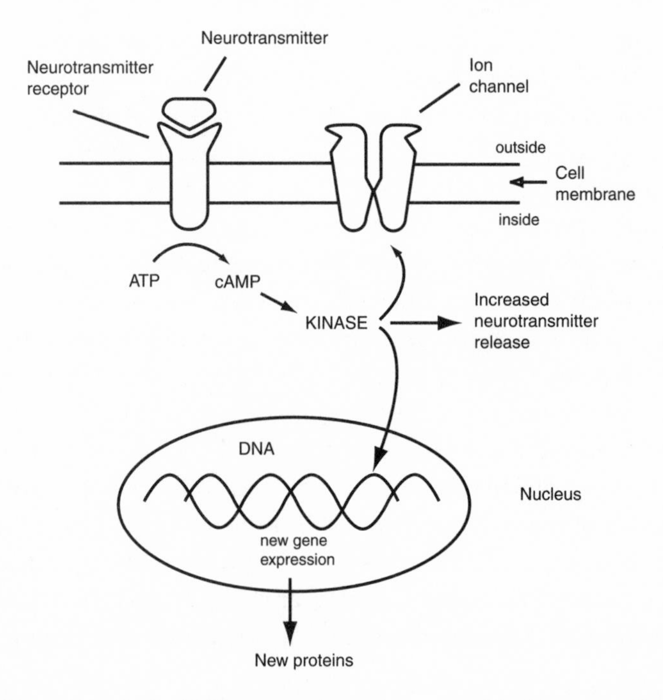

Figure 7.2 Cyclic AMP acts as a "second messenger" inside nerve cells.

term memory, one of the things the cAMP-activated kinase does is to cause (indirectly) the expression of new genes in the cell's nucleus. Some of these genes encode proteins that are much more brain-specific than the proteins involved in generating cAMP, such as those that take part in building more axons and dendrites. But the majority are not; they are just common, workaday proteins found in any cell of the body, and the changes brought about by and large simply make the cell work better and faster.

One of the most important revelations to come out of this work is that the transition from short- to long-term memory takes place within the very same cell. The development of memory consists not of transferring information from one type of neural circuit to another, where it is stored in a more stable form. Rather, the same circuit in which the information first arrived is altered in such a way that the next time the same information arrives, it is handled differently; usually, more rapidly and efficiently.

Facilitating interneurons are also involved in classical conditioning. As with *C. elegans*, Aplysia can learn to associate a signal coming in through one sensory pathway with information coming in through an entirely separate pathway. Facilitating neurons in a sense bridge the two response pathways. Facilitating interneurons and neurotransmitters such as serotonin are, as we see in the following chapters, implicated in a number of neurological reactions that occur in all animals, including humans. Other interneurons, using other neurotransmitters, tend to suppress in a long-term fashion, rather than enhance, the responses along certain pathways.

Learning and memory are at the very heart of all behaviors. Memory brings something new to behavior: Although behavior is physically acted out in the three dimensions of space, memory adds the fourth dimension of time. With memory, behavioral decisions can be made not just on the basis of information immediately confronting the organism, but also on the basis of information the organism encountered in the past.

As far as we can tell, the genes and proteins involved in implicit memory, from the simple nervous systems of the lowest multicellular organisms to the human brain, are remarkably similar. The reception of sensory signals and the conduction of nerve impulses, like the processing of information carried in DNA, is one of those phenomena that, once established, nature seems to have been reluctant to tamper with over evolutionary time. That is why the study of these molecular pathways, and the dissection of their underlying genetic regulation, is proving so valuable in understanding the neurological basis of human behavior.

Learning and Memory in the Common Fruit Fly

Drosophila melanogaster—the common fruit fly—hardly looks like one of the major work engines driving research in a central field like genetics. Swarming around the fruit bowl in the kitchen, or gathering around juicy melon rinds left behind on the breakfast table, they are about the same size as mosquitoes, and only slightly less annoying. Yet these tiny winged insects may have told us more about the genetic and molecular basis of life processes in animals than any other single model organism. Their hallowed place in the annals of genetics in this century can be traced back to the pioneering research of Thomas Hunt Morgan.

One of the fascinating things about Morgan is that while he originally rejected the hereditary theory of Mendel, and the evolutionary thinking of Darwin, he found through his own experiments, beginning in about 1905, that the theories of both were correct, and he ended up providing some of the most persuasive evidence in their favor. (He received the Darwin prize for his research in 1924, followed by a Nobel prize in 1933, for his work on heredity.) Morgan was among the first to realize that Mendel's theories about the inheritance of discrete phenotypic traits in plants also applied to animals. His work was a major factor in demonstrating that the genes responsible for phenotypes are physically associated with specific structures in the nucleus of every cell called chromosomes, and he especially contributed to the realization that every chromosome houses many different genes. His extensive genetic characterization of fruit flies rapidly made them the model organism of choice for other investigators looking at the genetic basis of a wide range of biological phenomena in animals.

Drosophila are both larger and more complex than the nematode roundworm *C. elegans*. Aside from a considerably more sophisticated motor system, which greatly enhances its food-gathering abilities, Drosophila also has a much more sophisticated nervous system. The nervous system of *C. elegans*, recall, has only 302 neurons; that of Aplysia 20,000. An adult Drosophila fly, tens of thousands of times smaller than Aplysia, has over 200,000 nerve cells. As in *C. elegans* and Aplysia, neurons in Drosophila can be grouped broadly into three types: sensory neurons, which pick up information about the environment; interneurons, which integrate and process this information and promote such things as learning and memory; and motor neurons, which direct the body's responses to environmental cues.

But whereas in *C. elegans* there are roughly similar numbers of neurons in each of the three categories just described, in Drosophila we see a trend also characteristic of evolving nervous systems: an increasing number and proportion of interneurons, devoted to the processing and integration of sensory input. In *C. elegans*, many of the processing interneurons are collected into a small structure in the neck region called a nerve ring, where the axons and dendrites of the neurons have extensive interconnections. This evolved into more sophisticated structures in Aplysia called ganglia, scattered throughout the body. There are similar ganglionic structures in Drosophila, such as the mushroom body, which also shows extensive interactions between nerve cells through their axons and dendrites. But these structures are proportionally much larger in Drosophila, and are clearly recognizable as part of what we would call a brain in higher organisms. The number of nerve fibers in the mushroom body increases during learning about the environment, and experiments have shown that damage to the mushroom bodies has a serious impact on learning ability in flies.

One of the first scientists to appreciate the advantages of Drosophila for the study of learning was Seymour Benzer. Like Sydney Brenner, Benzer is a distinguished molecular biologist whose early work was done in bacteria and the bacteriophages that infect them. Benzer was one of a small number of scientists who, independent of one another, had started out in physics but had been attracted to biology by the possibility of studying genetics in microorganisms.* They correctly fore-

* Other physicists who made outstanding contributions to the emerging field of molecular biology were Francis Crick, Salvatore Luria, Gunther Stent, and Max Delbrück. Like many others, they had been greatly stimulated by a book written in the 1940s by the Austrian physicist and Nobelist Erwin Schrödinger called *What Is Life?*, which held out the hope that ultimately biology could be understood in terms of the same laws that

saw the rapid advances that could be made working with life forms containing relatively few genes. But by the mid-1960s, a number of these researchers began thinking of ways to apply what they had learned about genetics to an understanding of how more complex, multicellular organisms operate; how they develop from a sperm and an egg; and why they behave as they do as adults.

Like Brenner, Benzer realized that molecular biology and genetics could be a powerful tool for dissecting complex behavior in higher organisms, and he was equally convinced that research would have to begin in a very simple animal system. But whereas Brenner almost single-handedly transformed *C. elegans* from a barely known soil worm to one of the more powerful research tools in molecular biology, Benzer decided to coopt the already well-established Drosophila model for his own studies. He made this decision after spending a year at the California Institute of Technology in 1965 exploring this possibility with Max Delbrück, among others, and joined the faculty a year or so later. Cal Tech had been Thomas Hunt Morgan's institution in his later years. Although Morgan, who died in 1945, never became comfortable with the concept of genes as chemical entities, his influence could still be felt in the strong genetics program that had evolved at Cal Tech. Ed Lewis, a student of Morgan's student Al Sturtevant, was just beginning his work at Cal Tech. (Lewis would later receive a Nobel prize for his own work in Drosophila developmental biology.) But perhaps the greatest factor in Benzer's choice of Drosophila was the enormous store of mutant fly stocks already available, many of them still at Cal Tech, and the ease of generating new mutations almost at will, thanks to the methodologies worked out by Morgan's successors over the years.

Shortly after moving to Cal Tech, Benzer began to explore learning behavior in flies, using a variant of the shock-avoidance learning protocol that was the basis for the associative learning tests described earlier in *C. elegans*. Flies are normally attracted to light. In a test chamber, flies were placed in a small tube with a light source at one end. A grid capable of imparting an electrical shock was placed at the lighted end of the tube, and the grid was coated with an odorant. Flies introduced

---

governed the rest of the physical universe, and that DNA would be written in a chemical language that could eventually be deciphered.

into these tubes immediately migrated toward the light, but as soon as they touched the grid they received a shock, and retreated from the light source. After several such "training sessions," the flies were placed in a tube with a light source and the odorant alone, with no electrical grid. They refused to migrate toward the light source, demonstrating they had learned to associate the odorant with light and an electrical shock. Interestingly, training regimens of this type result in greatly increased neuronal activity in the mushroom bodies of flies, underscoring the importance of this structure in the learning process.

Benzer then began screening existing Drosophila mutants for defects in this learning behavior. One of the more interesting mutations he and his students found came to be known as *dunce*. *Dunce* mutants were simply unable to learn the connection between the odorant and the unpleasant sensation of receiving an electrical shock. After the same number of training sessions, they still migrated toward the light in the presence of the repelling odorant. The mutants were examined carefully to be sure there was not some general physiological dysfunction that could explain their altered behavior. They had not lost their sensitivity to electrical pain; they disliked being shocked as much as normal flies. Separate tests showed they were also able to sense the odorant as well as normal flies. In fact there seemed to be nothing abnormal about dunce flies* except their inability to engage in associative learning.

It turns out that the defect in *dunce* mostly affects development of short-term memory, which of course indirectly eliminates long-term memory. If tested within a few minutes of the conditioning regimen, they show some learning ability (although greatly reduced compared with normal controls). But by thirty minutes, they seem to have forgotten everything; there is no transfer of even the little they have learned into long-term memory. Learning defects are also evident in positive reinforcement tests: *Dunce* mutants quickly forget when a particular odorant or other stimulus is associated with a food source they very much like. Development of long-term implicit memory in Drosophila involves the same kinds of changes in neurons that we saw in Aplysia. Extensive modification of interneuronal pathways is evident in the mushroom bodies of flies undergoing learning and mem-

* It would later turn out that females with two *dunce* genes are reproductively sterile.

ory acquisition; as we see below, the same cAMP pathways evident in Aplysia learning and memory also appear to be at play in Drosophila.

Pursuing the basis of the *dunce* learning defect in the context of the odorant/electrical grid system contributed important insights into the neural processes involved in learning, which underlies virtually every behavior in all animal species. Perhaps nowhere is this illustrated more clearly than in the role played by genes like *dunce* in the reproductive activities of flies—the most fundamental behavior of all.

Courting and Mating Behavior in Flies

The ritual is always the same. The male makes the first moves; he orients toward the female, approaches her, and uses his forelegs to tap her on the abdomen, signifying his interest. This initial display of interest depends in part on the male's vision, but to a larger extent on pheromonal signals released from the female's body surface; flies are able to mate successfully in complete darkness. Depending on her receptivity to the male's advances, she may simply ignore him—which is in effect a way of encouraging him—or she may poke back at him with one of her legs or wings, which is a way of discouraging him. But if he is a normal, healthy male, he will be obnoxiously persistent, tapping her and grasping her abdomen with both forelegs. He then begins playing a distinctive courting song by vibrating his wings at a particular pitch, and with a repeat of the chorus portion of the song, that the female is genetically programmed to recognize. There is no mistaking the male's intent: Although inaudible to the unaided human ear, in the vicinity of the female's antennae (where she picks up the vibrations from the song), the song is playing at nearly 100 decibels.

If the female is truly not interested in sex (for example, if she has recently mated with another male), she will release pheromones that repel the male, and in some Drosophila strains the previous male with whom she mated may have left male-repelling pheromones behind as well. Barring such active chemical and physical discouragement, the male will continue to play his courtship song, and then lick the female's genitalia, after which he tries to copulate with her. The female is relatively inactive in all this; if she is interested in mating, she basically slows her movements so that mating can occur. This immobility is actually an important part of encouraging the male to proceed. Mature males are attracted to and will try to mate with a completely immobilized (but pheromonally competent) female.

In Drosophila, as in most animal species, courtship and mating behavior does not vary under normal circumstances from one generation to the next. Yet the basic mating ritual can be shown not to be learned, leading to the conclusion that its fundamental elements are "hard-wired"—imprinted in the genes. That this is so can be demonstrated by the existence of many well-established genetic mutations that disrupt mating behavior. Some of the mutants that affect reproductive behavior are not very interesting, because they turn out to disable the entire organism, and can hardly be considered reproduction-specific. But other mutations are more informative, and none more so than *dunce*.

Normally, a male that has attempted to mate with a pregnant female, and has been seriously rebuffed, will not attempt to mate with pregnant females again for some time. Presumably the male learns that the repellent pheromones on a pregnant female are associated with uncooperative behavior, and he moves on to more promising territory. *Dunce* mutants are unable to learn this. Having been thoroughly repulsed by one pregnant female, they are as likely to try mating with another pregnant female (emitting the same repellant pheromones) as with a virgin female. In a population of only virgin females, *dunce* males do just as well as normal males in terms of courtship and mating success, showing that *dunce* does not affect any of the basic behaviors directly associated with courtship and mating. But in a population of mixed virgin and pregnant females (the "real world" for male flies), *dunce* mutants just don't get it—they spend a great deal of time in unproductive mating behavior, trying to impregnate already pregnant females. We can imagine that in the wild, after a few generations of such activity, the frequency of the *dunce* mutation in a population of flies would greatly decrease or be eliminated altogether.

The identity of the *dunce* gene in Drosophila was established in the early 1980s. The gene is located at one tip of the fly's X chromosome, and encodes an enzyme called phosphodiesterase (PDE) that is involved in regulating cellular levels of cAMP. This is the same cAMP signaling system that had been identified some years earlier in the acquisition of short- and long-term explicit memory in the nervous system of Aplysia. In Drosophila, as in Aplysia, there is certainly nothing brain-specific about cAMP per se. cAMP is a common "second messenger" used to help translate a signal arriving at a cell surface into an appropriate response by the cell.

One scenario for how cAMP may be involved in learning is that when mushroom body cells are stimulated by olfactory neurons that have received a pheromone signal, the level of cAMP rises inside the mushroom body neuron. The cAMP binds to and activates an enzyme called a cAMP-dependent kinase (Fig. 7.2), which may do several things. It may lead to the opening of an ion channel, triggering the cell to depolarize and pass a signal on to other neurons to which it is connected. It may also be involved in the induction of new proteins in the cell, which alter the cell's synaptic activity. Proper functioning of the neuron, however, depends on the prompt removal of cAMP and deactivation of the kinase, so that the ion channel closes, and the neuron can repolarize. Removal of cAMP is the specific task of PDE. Drosophila has not one but three different PDE enzymes, each encoded by a different gene. The gene encoding the form of PDE affected by the *dunce* mutation is expressed preferentially in nerve cells, and perhaps specifically in mushroom body neurons.

Studies with several other Drosophila learning mutants also turned up genes involved in cAMP metabolism. A mutation called *rutabaga* results in an inability to make cAMP in the first place. *Rutabaga* flies show almost exactly the same learning defects as *dunce*, and the *rutabaga* gene, like the *dunce* gene, is preferentially expressed in mushroom body cells. *Rutabaga* mutants show that the generation of cAMP in response to a signal, as well as its timely removal, is important in learning. The learning mutant *amnesiac* appears to be defective in a gene encoding a neuropeptide-like protein that activates cAMP in cells.

The Molecular Basis of Learning and Memory in Mammals

Mammals are fully equipped with the implicit memory function characteristic of lower organisms like Drosophila and Aplysia; it is readily demonstrable even in human beings. Habituation of reflexive responses can be seen in the response of someone to a sudden loud noise; he or she will show a standard reflexive startle response to the first sound exposure, but upon repeated exposure to the same noise will gradually stop responding. After an appropriate interval without the noise bursts, a normal reflexive reaction is restored. Classical conditioning of reflexive responses can be elicited in humans as well. One of the most commonly used tests for conditioning of implicit memory is the eye blink test. A small jet of air directed against the cornea will cause a human to blink reflexively. If the jet of air is repeatedly preceded by a sound tone delivered through a headset, after a number of trials the sound tone alone will elicit a blinking response.

Many of the neurons participating in implicit memory in mammals can be found in a region of the brain called the cerebellum (Fig. 7.3). Although accounting for less than 10 percent of the human brain's volume, more than half of all of the brain's hundred billion or so neurons are located there, with extensive projections to all parts of the brain and spinal cord. The cerebellum is a relatively ancient region of the vertebrate brain, and plays a major role in muscular coordination and balance, among other things. Since implicit memory involves almost exclusively the coordination of motor functions, it is perhaps not surprising that the cerebellum plays such a large role in this form of memory. Humans and other mammals with damage to certain regions of the cerebellum are unable to undergo conditioning of the eye blink response, for example.

There is also a tiny, paired structure in the brain called the amygdala that lies near the hippocampus (Fig. 7.3). The amygdala participates in implicit memory-linked behaviors in mammals such as conditioned responses, and is also thought to be the seat of something called "emotional memory." It has been best characterized in both animals and humans as the site where implicit memories involving an element of fear are stored. Evoking these memories often sets in motion a series of physiological changes preparing the organism to react in a strongly defensive manner. But damage to this area in humans can result in a general impairment of the ability to interpret emotions in oneself and in others. It is thus possible that a major carry-forward of implicit memory in humans is related to emotionally based behavior.

The importance of cAMP in the formation of long-term implicit memory in lower organisms was initially surprising to scientists, and stimulated them to look for a role of cAMP and cAMP-dependent enzymes, such as PDE, in learning and memory formation in mam-

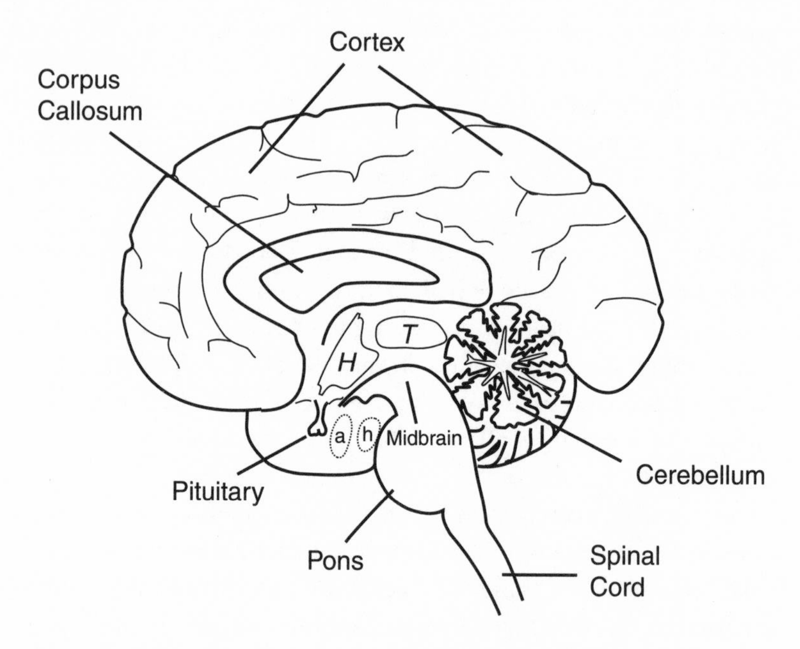

Figure 7.3 The human brain. H: hypothalamus; T: thalamus; a: amygdala; h: hippocampus.

mals, including humans. And in fact a *dunce*-type PDE gene was isolated from rats in 1989. A copy of the Drosophila *dunce* PDE gene was used to "fish out" the rat gene from a preparation of rat DNA. If the same genes from two different sources are sufficiently similar in terms of their sequence, they will adhere specifically ("hybridize") to one another; the Drosophila gene was thus used as "bait" to probe rat DNA for a similar gene. The fact that the experiment worked tells us that the *dunce*-type PDE gene has likely been conserved through the evolutionary period separating Drosophila from rats—a period of at least 600 million years. By comparing the sequence of the *dunce* gene in the two species it can be concluded that the sequences of the respective encoded PDE proteins are 75 percent identical, which is indeed an extraordinary degree of conservation over such a long period of time.

It turns out that rats have seven different PDE genes (PDE 1–7), only one of which (PDE-4) is preferentially expressed in the nervous system like the Drosophila *dunce* gene. When PDE-4 function was disrupted in rats, cAMP was elevated preferentially in nerve cells, suggesting that rat PDE-4 plays a role similar to PDE in flies. There is also functional conservation; the rat gene, when introduced into *dunce* flies, was able to correct partially the learning defect in these flies!

Cyclic AMP may well regulate memory formation in rats. The rat PDE-4 enzyme can be selectively inhibited in normal rats with a drug called rolipram. If rats are treated with a substance called scopolamine, or subjected to a mild stroke or certain kinds of electrical shock, they suffer impaired transfer of information from learning to long-term memory. But this deficit can be reversed by rolipram. Isolated rat nerve cells treated with rolipram show numerous deficits, such as an inability to release neurotransmitters. One of the effects of rolipram in these rats is to increase the levels of cAMP in neurons involved in the learning and memory pathways. This suggests that one of the impairments in rats treated by scopolamine, stroke, or electroconvulsive shock may be a lowering of cAMP levels in neural pathways. Such rats are similar to the *rutabaga* mutants described earlier in Drosophilia, demonstrating that not only too much but too little cAMP in neurons can impair learning and memory.

The first human PDE-4 gene corresponding to Drosophila *dunce* was fished out of human DNA in 1990, using as bait a copy of the rat *dunce* gene. As in the rat, there are seven different PDE genes in humans, and the human PDE-4 gene appears to function essentially like its fly and rat counterparts. The PDE-4 protein in humans is, like that of the rat, inhibited by the drug rolipram. This is of great interest because rolipram is an effective antidepressant drug in humans. One of the things rolipram and many other antidepressant drugs do is to increase the level of a special protein in nerve cells that binds cAMP, indicating at least indirectly that rolipram-treated cells may have elevated cAMP levels. The cAMP-binding proteins induce the expression of other genes in nerve cells, genes that are suspected to be involved in the specific response of the cell to incoming signals. It is intriguing that this system is also involved in a form of mental functioning in humans.

Humans clearly have explicit memory, which requires a conscious effort to retrieve previous information, including visual images, stored in connection with previous experience. Memories of people, places, and events in our lives are all forms of explicit memory, and they are formed in a separate region of the brain—the hippocampus (Fig. 7.3).

Because of the physical separation within the brain of these two forms of memory, it is possible to lose implicit memory (in certain types of brain damage, for example) without affecting explicit memory, and vice-versa. Unlike implicit memory, which requires a number of repeats of the learned event in order to be converted from short- to long-term memory, long-term explicit memories can be generated from a single learned or recorded event.

So what is the *dunce* gene doing in mammals? Is it involved only in implicit memory, or might it also be involved with the formation of long-term explicit memory? When researchers looked to see where in the brain mammalian *dunce* is expressed, it was found both in the cerebellum—one site of implicit memory formation—and in the hippocampus, where explicit memory predominates. In Aplysia, as we saw, long-term implicit memory formation involves a process called facilitation, wherein special interneurons act to strengthen the synaptic connections in sensory pathways that have been stimulated during learning. A functionally similar process in long-term explicit memory formation is called "long-term potentiation." As a result of long-term memory formation, increasing numbers of synapses are formed in the hippocampus, and at least some of these synaptic connections are dependent upon cAMP elevation inside the hippocampal neurons. Moreover, one of the changes brought about in neurons in long-term potentiation is increased output of neurotransmitters in response to an incoming signal. Hippocampal cells that have been placed into cell culture will release neurotransmitters when appropriately stimulated. If the cells are treated with drugs such as serotonin that elevate intracellular levels of cAMP, the output of neurotransmitters is greatly increased.

In addition to cAMP-generating and -degrading enzymes, another fairly common enzyme with the rather clumsy name "calcium-calmodulin-dependent kinase II" has been shown to be induced during learning regimens, and to be involved in long-term explicit memory formation. (We follow the lead of scientists who work with this enzyme and call it simply CaMKII, pronounced "cam-K-2.") CaMKII is analogous to the kinase that is activated by cAMP, except that it is activated by Ca^{++} rather than cAMP. This is the very same calmodulin molecule we saw in Paramecia. In mammalian cells, just as in Paramecia, calcium entering a cell is bound by calmodulin, and the

Ca^{++}–calmodulin complex can in turn activate various proteins within the cell. Like the cAMP-dependent kinase, the Ca^{++}-activated CaMKII acts to induce the synthesis of new proteins involved in creating and strengthening synaptic connections. Like cAMP kinases, calcium–calmodulin kinases, including CaMKII, are common in cells throughout the body. In the brain, CaMKII is particularly concentrated in the pre- and postsynaptic regions of the hippocampus, and in regions of the forebrain.

Recently a mouse lacking the CaMKII was generated in the laboratory. Although otherwise normal, these mice were markedly impaired in the development of long-term, hippocampus-localized explicit memory. They were slower to memorize, in relation to other visible markers, the location of a platform submerged just below water in a pool in which they were forced to swim to safety. In a separate set of experiments, they were also slower to memorize the pathway through a maze which allowed them to escape from bright light, which they normally avoid. More recently, it has been shown that some implicit memory pathways in mammals may also utilize CaMKII. And when scientists then went back and disrupted CaMKII function in Drosophila, the ability to learn courtship and mating behavior (which involves implicit memory) was severely impaired. This is an excellent example of the high degree of evolutionary conservation of cell signaling mechanisms, from Paramecia through mammalian nerve cells.

Learning and memory will certainly be quantitative traits, involving many more genes than just PDE-4 and CaMKII. But if these genes are just the tip of the iceberg, in terms of the neurological pathways involved in learning and memory, they are still a part of the iceberg. Other genes that affect learning and memory have already been discovered in the same fashion. *Rutabaga*, like *dunce*, involves a type of gene commonly found in cells throughout the body. Another mutation affecting learning called *shaker* involves a gene encoding a potassium channel; defective function of potassium channels was behind some of the behavioral mutations we saw in Paramecia in Chapter 2. Potassium channels are by no means unique to brain cells; they are expressed by many different cells throughout the body.

There are means other than simply tracking mutations for identifying genes involved in behavioral pathways such as learning and memory, and we examine some of these strategies in the chapters that follow. One of the goals of behavioral genetics is to identify at least the major genes involved in quantitative behavioral traits, and to understand how allelic variations in these genes contribute to overall behavioral patterns. The finding that many of the genes involved in rather specific behaviors are not at all behavior-specific genes would have surprised many of the scientists who set out on the original quest to define the genetic basis of behavior. But no longer–this finding fits in quite well with what we have learned about the genetic basis of a great many other biological processes in the past thirty years or so.

Finally, it is important to note that when we speak of learning and memory, whether in worms or in human beings, what we really are talking about is the impact of the environment on the nervous system. It is here that genetics and the environment become inextricably entangled. Genetic modulation of the nervous system affects how information from the environment is perceived, and also how the organism responds to that information. But at the same time, when we speak of new neuronal connections and interactions generated as a result of learning, or when we speak of synaptic alterations emerging as a result of experience, we are really talking about the impact of the *environment* on the nervous system, and ultimately the impact of the *environment* on behavior. We must not lose sight of this important fact as we begin our exploration of the genetic basis of human behavior in the chapters that follow.

8

The Role of Neurotransmitters in Human Behavior

In the preceding chapters, we focused largely on the role of genetic variation in explaining animal behavioral variability. In those experiments designed to explore the role of genes in animal behavior, the laboratory "environment" was held as constant as possible, in order to show the effects of genes more clearly. In the remaining chapters, we turn our attention to the variability we observe in human behavior. We again ask to what extent this variability can be explained by genetic variation. But this is the point at which we must also begin to explore in more depth the extent to which environment can influence variations in behavioral responses. The environment in which human beings live and work, love and learn, cannot be so readily controlled. Humans do not live in laboratories, and the effects of environment on their behavior cannot so easily be set aside to reveal underlying genetic influences.

Everything we know about both animal and human behavior suggests it is ultimately an expression of events guided by the nervous system, aided to varying degrees in more complex multicellular animals by some of the hormones of the endocrine system. As decision-making neurological processes became increasingly concentrated in a centralized brain over evolutionary time, behavior became essentially an expression of the brain's response to information coming to it from both inside and outside the body. It is through the sensory elements of the nervous system that the environment affects us, a point we must keep in mind as we evaluate the relative roles of genes and environment in behavior. Genetic variation may affect not only our responses to the environment, but our very perception of it in the first place.

So it is to the nervous system we look for the expression of genes whose variability within a population could explain at least in part the variability in behavior we see among human beings. The many genes that direct activities inside a neuron, such as those currently being defined in Aplysia and other experimental models—particularly genes involved in signal transduction—certainly make up one important class of genes to consider. But an equally attractive set of genes for use in probing behavior includes those genes responsible for the production, regulation, and function of the various neurotransmitters and their receptors.

Neurotransmitters are the way nerve cells communicate with each other, and with other cells in the body. We use neurotransmitters to relay information about the environment to the brain, to analyze that information, and to set in motion appropriate bodily responses. The human brain consists mostly of interneurons that form bridges between cells in different parts of the brain. The resulting chemical cross-talk is necessary if the brain is to draw on all of its stored information in formulating responses to internal and external events. Interneurons connect nerve cells in all regions of the brain, from those regions in which experiences are first perceived, through extensive switching systems that route incoming signals through memory banks, and through the intellectual centers of the brain's cortex. The brain is literally bathed in information flowing in the form of neurotransmitters. Nerve cells in the brain and in the spinal cord also use these same neurotransmitters to communicate with other cells in the body, for example, to instruct a gland cell to release a hormone, or a muscle cell to contract in order to initiate movement.

Before we look at how neurotransmitters influence behavior, let's take a look at how neurotransmitters work at the level of individual neurons, because this will help us understand not only how communication between nerve cells takes place, but also how certain drugs can sometimes be effective in modulating behavior.

Neurotransmitters come in two forms; small protein molecules called neuropeptides, or even smaller chemical structures like acetylcholine and serotonin (Table 8.1). Among the peptide neurotransmitters are molecules such as β-endorphin, which mimics morphine in managing pain in the body. But by and large it seems to be the small molecular neurotransmitters that carry the heaviest burden in regulating brain activity affecting behavior. Of the nine small molecular neurotransmitters listed in the table, eight either are amino acids—the individual building blocks of proteins—or are derived from amino acids. (Acetylcholine is the exception.)

In carrying out their various tasks, neurotransmitters do not act *within* cells; they act *between* cells. Each neurotransmitter is synthesized within the neuron itself, and stored in small releasing granules at the axonic tip (Fig. 8.1). When a nerve cell is depolarized, it releases the contents of these granules into the synaptic cleft—a tiny, open space between the presynaptic axon of one cell and the postsynaptic dendrite of another. The neurotransmitters move across the synapse by simple diffusion and bind to receptors on the postsynaptic, dendritic tip of the receiving neuron. All neurotransmitter receptors are proteins embedded in the surface membranes of neurons. The receptors are

Table 8.1. *Neurotransmitters*

Small molecules	Some common neuropeptides
Acetylcholine	β-Endorphin
γ-Aminobutyric acid	Neuropeptide Y
Dopamine	Neurotensin
Epinephrine	Oxytocin
Glutamate	Prolactin
Glycine	Somatostatin
Histamine	Substance P
Norepinephrine	Thyrotropin
Serotonin	Vasopressin

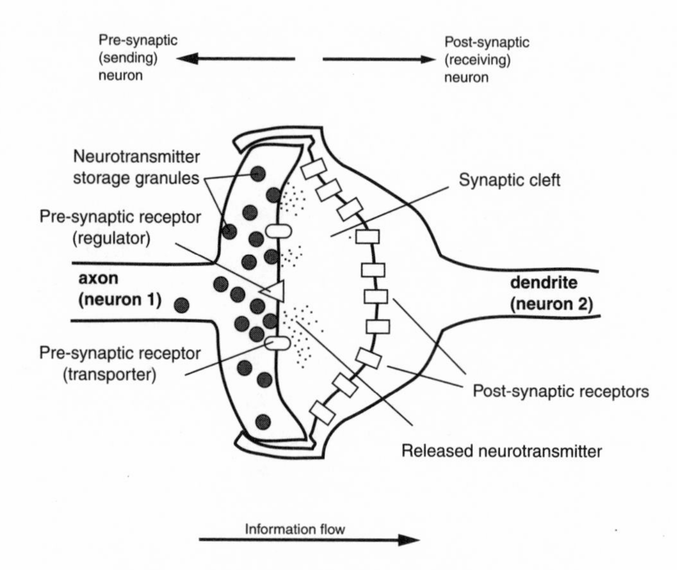

Figure 8.1 Detailed view of a neurological synapse.

either themselves ion channels, or they are intimately connected to ion channels. If an appropriate receptor is not present on the postsynaptic fiber, nothing happens. If a receptor is present, several things may happen. If the receiving cell is another neuron, the incoming signal will act either to facilitate or to suppress release of that neuron's own transmitter. If the receiving cell is a muscle cell, it will contract.

Although there is only a single form of each neurotransmitter found in the body, there may be as many as a dozen or more different kinds of protein receptors for each neurotransmitter. Different neurons, in different regions of the brain and carrying out different functions, may express different receptors for the same neurotransmitter. This allows the same neurotransmitter to affect neurons in different ways, depending on the type of receptor they display; each receptor, when occupied, triggers a different kind of reaction within the receiving neuron.

Most neurons also have receptors for the neurotransmitter they are programmed to release, and these "presynaptic" receptors play important roles in nerve function. These receptors are also common targets for a number of behavior-modifying drugs, as we see below. One category of presynaptic receptor is referred to as a transporter, or sometimes as a reuptake receptor. This receptor is located at the tips of neuronal axons, and is thus part of the architecture of synapses formed by two neurons. When a neurotransmitter is released from an axon into a synapse, some of it is taken up by the receiving dendrite (i.e., postsynaptically). But a good deal of the released neurotransmitter remains unused and lies idle in the synapse. This leftover neurotransmitter would be mixed with the next burst of neurotransmitter released by the same neuron when it fires again, and eventually unused neurotransmitter would build up to intolerable levels and begin to interfere with normal synaptic function. This is where the presynaptic transporters come in. To clear the synapse of neurotransmitter from the previous burst, the transporter resorbs the unused neurotransmitter back into the releasing neuron, where it is either recycled to storage granules or destroyed by special enzymes.

Another type of presynaptic receptor acts to regulate neurotransmitter release. These regulatory receptors do not function in neurotransmitter reuptake, but if they detect too much neurotransmitter in the immediate region of a neuron, they may shut down further neurotransmitter release until the situation normalizes. This can be readily demonstrated with nerve cells grown in culture. These cells can be stimulated by various means to release neurotransmitter into the culture dish, but if enough of the neurotransmitter is added to the culture fluid before stimulation, the neurons are unable to "fire," to release their own neurotransmitter.

Each neuron can be connected through its single axon to dozens of other neurons, and may itself receive input from hundreds of other neurons through its many dendrites (see Fig. 4.2). Each neuron is capable of releasing only a single kind of small molecular neurotransmitter from the tips of each of its axonic projections, although it may release several kinds of neuropeptides. Individual neurons normally have receptors for a wide range of different neurotransmitters at the tips of their dendrites. A given neuron can thus be influenced by many different types of other neurons to which its dendrites are connected. Communication among neurons in the brain takes place via an orchestrated release and uptake of varying kinds of neurotransmitters between and among cells. The action of a given neuron will depend on the sum of the chemical inputs it is receiving at any given time from other neurons.

Measuring neurotransmitter levels in something as complex as a brain presents certain difficulties. In many animal studies, it is possible to insert measuring devices directly into reasonably well-defined subregions of the brain, so that information about local concentrations of a given neurotransmitter can be obtained. Equally often, however, levels are simply measured for the brain as a whole. In humans we are even more restricted; neurotransmitter levels can only be guessed at by measuring the neurotransmitter or one of its metabolites in spinal fluid, blood, or urine. These estimates are of necessity approximations, and even the more precise measurements in animal brains do not tell us what changes may be taking place along individual nerve tracts. Nevertheless, some very striking correlations between neurotransmitter levels and behavior have been made in both animals and humans, and we explore some of these in the following chapters.

All neurotransmitters play some role in behavior, including the neuropeptides. Nevertheless, the neurotransmitters most commonly implicated in behavior modulation are the small molecular transmitters norepinephrine, dopamine, and serotonin. In the sections that follow, we take a brief look at how these neurotransmitters play a role in three important forms of modulating behaviors in humans: impulsivity, and psychological depression, and learning.

Impulsivity

Impulsive behavior is defined as any behavior which, by reason of direct experience or absorbed knowledge, individuals may reasonably know is not in their best interest, but which they consciously pursue anyway. It is often characterized as action without sufficient reflection. Some of the more classically recognized forms of impulsive behavior, such as sudden and seemingly unprovoked aggression toward persons or objects, drug addiction and certain eating disorders, pyromania, and repeated suicide attempts, are listed in Table 8.2. To a large extent, impulsiveness overlaps with the introversion–extroversion axis discussed in Chapter 1, but impulsivity usually has a negative connotation because of the harm it can cause not only to the impulsive individual but, through that individual's behavior, to others. Impulsive acts are often preceded by a period of rising tension, which resolves into a sense of relief and well-being once the act is completed. To a large extent, all of this is regulated by neurotransmitters.

Table 8.2. *Behaviors with an underlying element of impulsiveness*

Eating disorders	**Gambling**
Hyperphagia (overeating)	
Anorexia nervosa	**Kleptomania**
Bulimia nervosa	
	Paraphilia
Spontaneous aggression	Voyeurism
	Sadomasochism
Substance abuse	Exhibitionism
Alcohol abuse	
Drug abuse	**Suicide attempts**
Sensation seeking	**Attention-deficit hyperactivity disorder**
Risk taking	
Thrill seeking	
	Trichotillomania (a compulsive pulling out of one's own hair)

In its favor, we should point out that impulsivity has also been implicated in some forms of human creativity, both artistic and scientific, and thus in some situations it can have a positive as well as a negative outcome. Risk taking, for example, can be a positive trait under some circumstances—think of people who break with everything around them and set out to explore the world, or move to a completely different culture, or go to the moon. Think of the person of modest means who risks everything to start a new software business in the garage. But while sometimes enriching or rewarding in some areas of human endeavor, impulsive forms of behavior are more often disadvantageous, both to the individual and to society. Impulsive behaviors that lead to a positive outcome are sometimes referred to as "functional impulsivity," while those that result in harm to the individual or to others are called "dysfunctional impulsivity."

Impulsive individuals are believed to have a low degree of what psychologists refer to as "cortical arousal"—the involvement of the cortex, with its rich stores of information gleaned through education and experience, in the decision-making process. Behavioral responses begin in a relatively primitive region of the brain called the midbrain; in fact, many behaviors can be completely orchestrated and implemented from the midbrain without involving the cortex. We don't need cortical involvement in every decision we make; if a child steps out from behind a car into the street in front of the car we are driving, we do not want to spend much time thinking about what to do. But before deciding on that one last drink for the road, or verbally or physically assaulting the bartender for refusing to give it to us, a little reflection might be a good thing. And that is where impulsive people have a problem.

One of the principal means by which neurons of the midbrain wake up the cortex and get it involved in decision making is through the neurotransmitter serotonin. Serotonin is made in the body by chemically modifying a single amino acid called tryptophan. Tryptophan is one of a small number of "essential" amino acids that we cannot make in our own body, but must take in through the food we eat. In humans, serotonin was originally described as a natural substance in the body causing the smooth muscle around blood vessels to contract, which constricts the blood vessels and increases blood pressure. Serotonin also causes smooth muscle along the intestinal wall to contract, aiding in the movement of food along the gut. It was only in the 1950s when researchers, looking at its distribution in the body, finally detected serotonin in the brain and began to suspect it might play a role in neural function. Only about 1 percent of the body's total serotonin is found in the brain. Serotonin made elsewhere in the body cannot be used in the brain, because serotonin cannot cross the blood–brain barrier. However, dietary tryptophan can cross into the brain, and it is used by nerve cells to synthesize their own serotonin.

Serotonin has an unusually large number of different receptors present on various cells in the brain; a total of fifteen receptors, spread across seven different structural classes. Most of these are postsynaptic receptors that aid in the transmission of serotonin-based nerve impulses. There are two presynaptic receptors that are important in behavior: the serotonin transporter and a regulatory receptor referred to as the serotonin-1β receptor. Both of these receptors play an important role in regulating serotonin levels in the brain. Inhibitors of the serotonin transporter cause a buildup of serotonin in synapses, a useful strategy in the treatment of a number of problems associated with too little serotonin, such as psychological depression, which we will discuss shortly. The serotonin-1β regulatory receptor senses the level of serotonin in the immediate vicinity of the releasing cell, including the synapse, and regulates the cell's ability both to produce and to release more serotonin. This receptor, too, is a useful target for drugs intended to modulate serotonin levels.

Serotonin is involved in several types of communication in the brain. It is particularly used during periods of self-directed behavior: Activities such as eating, grooming, or simply resting and thinking are accompanied by high levels of brain serotonin. But it is also the main chemical messenger used to wake up the cortex. In this role, serotonin is used largely by neurons found in the so-called "raphe nucleus" of the midbrain region, where possible behavioral responses to various stimuli are first formulated (Fig. 8.2). Neurons from the midbrain send axonic projections into the frontal cortex and temporal lobes, where they release serotonin, which acts on neurons displaying serotonin receptors. These neurons in turn become involved in the further processing and interpretation of the incoming messages, and in selecting an appropriate response. Without this critical input from the cortex,

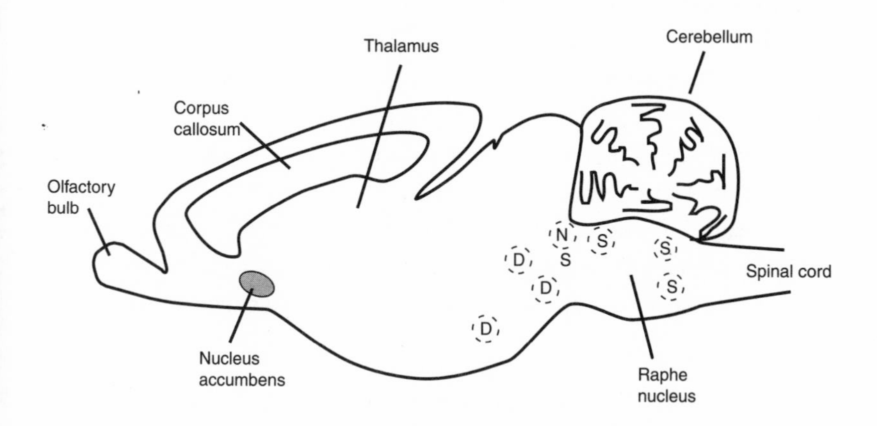

Figure 8.2 Detailed view of the midbrain. (See Figure 7.3 for orientation.) The locations are shown for clusters of neurons releasing norepinephrine (N), serotonin (S), and dopamine (D).

which includes reference to stored memories based on previous learning and experience, many behavioral responses remain largely midbrain-controlled; they are rushed, and may not always be optimal for the situation at hand.

Another neurotransmitter that plays a role in impulsive behavior is dopamine. Like serotonin, dopamine is made by modifying a simple amino acid, in this case tyrosine. Unlike tryptophan, the precursor of serotonin, tyrosine cannot cross the blood–brain barrier, so synthesis of dopamine actually begins in the liver, where tyrosine is converted to a molecule known as L-DOPA. This molecule is then transported through the bloodstream to the brain, where it is used by nerve cells to make dopamine.*

The role of dopamine in impulsive behavior is most likely tied to its involvement in the brain's system for rewarding particular behaviors. Certain elemental types of behavior critical to the survival and reproduction of animals, such as eating food, drinking water, or mating, produce a sense of satisfaction or "rightness" that actively reinforces engagement in these behaviors. Inducing these sensations is the job of dopamine. The neurons controlling reward are also found in the midbrain, just forward of the raphe nucleus (Fig. 8.2). These neurons send axons into the nucleus accumbens area, where release of dopamine promotes the sense of well-being. Impulsive individuals almost always experience a feeling of satisfaction or release just after completion of impulsive acts, most likely related to the release of stress and tension that often precedes such acts. Such feelings are often reported, for example, by persons who gamble uncontrollably, and even by those who steal things repeatedly. Relief of stress is something that the body constantly strives for, and so the involvement of dopamine in rewarding and reinforcing behaviors that reduce stress makes biological sense. As we see in Chapter 11, all addictive drugs, such as cocaine, alcohol, and heroin—and even common stimulants such as caffeine and nicotine—achieve their euphoric effects at least in part by increasing

---

* Dopamine is also involved in muscular coordination in a region of the brain called the substantia nigra. Persons with Parkinson's disease are often treated with L-DOPA to increase dopamine levels lost when certain midbrain neurons die. This has been described clearly and poignantly in Oliver Sacks's book *Awakenings* (New York: P. Smith, 1990.)

dopamine levels in the brain, and this is clearly a major factor in reinforcing the underlying substance-abusing behavior. They do this either by increasing dopamine release or blocking its reuptake at synapses.

Like serotonin, dopamine interacts with multiple postsynaptic receptors. These receptors are not scattered randomly throughout the brain; each tends to be localized in specific regions of the brain, and is assumed to promote slightly different responses in the neuron it serves. There is also a presynaptic transporter for dopamine. This receptor is a target for certain stimulatory drugs such as cocaine or amphetamines, and drugs that interfere with access to this receptor can often attenuate the drive toward consumption of such drugs.

A number of studies have suggested that impulsive behavior in humans is highly heritable, and that genetics is at least as important a factor as environment in determining whether a person is impulsive. These conclusions are drawn from family and adoption studies and, most important, from studies of monozygotic and dizygotic twins, reared together and apart. As with most other personality measures, shared environmental influences—essentially, the home rearing environment—have a minimal impact on whether someone is or is not impulsive. We look more closely at the data for the heritability of impulsivity, and possible roles of environment, when we discuss specific impulsive behaviors in more detail in subsequent chapters.

Neurotransmitters and Depression

Precise clinical definitions of the currently recognized forms of depression can be found in every psychiatrist's standard handbook: the *Diagnostic and Statistical Manual of Mental Disorders*, published by the American Psychiatric Association. Depression can take many forms. Two major divisions are unipolar and bipolar depression. In unipolar depression, individuals are simply depressed for varying lengths of time. In bipolar disorder, bouts of depression are interspersed with periods of intense activity or mania. Unipolar depression can be relatively transient. Many of us have experienced at least mild depression at one time or another, triggered by external events such as loss of a loved one, failure in a personal relationship, or stress at work. Most of us pass through this state after a period of time, and resume our normal activities. We look outward again at the world, and join in its challenges and rewards, rather than continue to trace endlessly the internal pathways of our underlying despair.

But some individuals are chronically depressed, experiencing deep, unshakable sadness, greatly diminished interest in their surroundings, and debilitating loss of self-esteem. They experience physical as well as mental lethargy, appetite and sleep disorders, perturbed sexual function, a sense of utter hopelessness, and an inability to engage meaningfully with those around them. When these symptoms persist for long periods of time, and interfere with an individual's ability to function in a family or societal context, the diagnosis is referred to formally as major depression. Individuals with major depression simply cannot crawl back up and out of the hole they have fallen into without some sort of help.

Estimates of the number of people who have experienced prolonged periods of depression in their lifetime run as high as 10 percent. For reasons that are not understood, women are two to three times more likely to experience major depression than are men. As many as one in five individuals with major depression will commit suicide. In this regard, men outdo women; while three times more women than men with major depression attempt suicide, four times more men succeed. The attempt to take one's own life is, on one level, certainly a reflection of the impaired judgment that accompanies a severely depressed state. However, the pain of chronic, severe depression is no less real than the pain of a chronic and severe physical disease such as cancer, and on that level suicide may well seem a perfectly rational way to end the suffering. But depression has an organic basis in the biochemistry of the mind, and it can be treated—in the vast majority of cases, quite successfully.

Like impulsivity, depression has a definite genetic component as shown by family, twin, and adoption studies. Two large studies have examined the occurrence of major depression in monozygotic and dizygotic twin pairs. One recently reported study included both male and female twin pairs. This study found that the likelihood that both twins would experience major depression at least once during their lifetime was about twice as high for monozygotic twins as for dizygotic twins. This was true for both male and female twin pairs. The heritability of major depression was estimated at 39 percent for both sexes. A second large study, which followed male twin pairs from the Vietnam War era, estimated heritability at 36 percent. The findings from twin studies are supported by adoption studies: Adopted children whose biological parents have experienced chronic depression will themselves be highly susceptible to depression even though their adoptive families have no history of depression.

But like all other human personality traits, it is a heightened susceptibility to depression that is heritable, not depression per se. Environmental factors, in particular stress, play an important role in converting a simple predisposition into the nightmare reality of clinical depression. The twin studies have made it clear, however, that it is nonshared, rather than shared, environmental experiences that are most important in the development of depression.

The notion that depression has an underlying chemical basis, and therefore should be amenable to pharmacological as well as psychotherapeutic intervention, has been around for some time. It has been known since the 1960s that depressed individuals nearly always have low levels of serotonin in their central nervous systems. This is thought to result in a failure to fully arouse the cerebral cortex to participate in an evaluation of what is happening to the individual internally. This could provide a more reasoned appraisal of one's situation and avoid the sense of helplessness and hopelessness associated with major depression. Among the most effective chemical treatments for depression are drugs that block the serotonin transporter in humans, effectively increasing the concentration of this neurotransmitter within synapses along serotonin pathways in the brain. These drugs, referred to collectively as SSRIs (selective serotonin reuptake inhibitors), include fluoxetine (Prozac), paroxetine (Paxil), and sertraline (Zoloft). However, the connection between serotonin levels and states such as depression may be complex. Overall levels of serotonin in the brain rise almost immediately after administration of SSRIs, yet an impact of these drugs on behavioral states often takes many days.

Alterations of dopamine function can also contribute to depression. A subset of depressed individuals appear to have lower than normal levels of dopamine, and drugs that elevate dopamine levels—transporter inhibitors, for example—are effective in treating this subset of patients. Depressed individuals often report loss of a sense of pleasure in many things that once brought them joy. Given the known role of dopamine in psychological reward systems, it is not surprising that impaired dopamine function could contribute to depression.

Another neurotransmitter that plays a role in depression is norepinephrine, which is made by further chemical modification of dopamine. Like serotonin, norepinephrine plays a role in brain arousal, but it does so in a slightly different sense. Norepinephrine alerts the brain to the presence of novel and potentially threatening events in the external environment. When a given set of sensory neurons is stimulated by a loud noise, for example, or an unexpected visual image or strange smell, norepinephrine-releasing neurons arouse other neurons all over the brain to a higher state of alert. The animal becomes more wary and vigilant. The neurons involved in this function are again located in the midbrain region, just under the cerebellum (Fig. 8.2). They send axonic projections to many different brain sites and to the spinal cord. These neurons may very well be the mammalian analog of the Aplysia facilitating interneurons discussed in the previous chapter. When these neurons are crippled either surgically or chemically, animals are much slower to respond to external danger. They also spend a good deal more time sleeping, and are hard to arouse. Norepinephrine release is also one of the brain chemicals controlled by our biological clocks to regulate sleep/wake cycles. Norepinephrine levels normally peak in the body at around noon, reaching a low point in the middle of the night.

Norepinephrine is different from dopamine and serotonin in that it also plays a role in rousing not just the brain, but the rest of the body as well, to action in a crisis situation. Norepinephrine is used by neurons in the so-called "autonomic nervous system," which connects the brain with the body's system of organs, glands, and blood vessels. The autonomic nervous system consists of a series of nerve fibers and associated ganglia that run along either side of the spinal column. These fibers are connected to the midbrain through the brainstem, providing a major pathway for communication between the brain and the body. This system is called autonomic because it controls functions that are essentially automatic, not requiring thought or judgment—such as heartbeat, breathing, blood pressure, and movement of food through the gut.

Norepinephrine is a key neurotransmitter used by the brain to communicate with the rest of the body throughout the autonomic system. Norepinephrine can either heighten or repress the sensitivity of a targeted neuron, depending on the type of receptor that neuron displays.

For example, in a stressful situation, autonomic neurons leading to the heart release norepinephrine, which binds to activating receptors on heart muscle, increasing both the rate and strength of heart muscle contractions. Autonomic nerve fibers attaching to smooth muscle surrounding blood vessels also release norepinephrine, which again binds to activating receptors and results in increased blood pressure. These receptors interact with the cAMP-generating systems within the targeted cells. When norepinephrine binds to a repressing receptor on a cell, it decreases cAMP levels in that cell, generally dampening cellular activity. Binding of norepinephrine to activating receptors has the opposite effect; it increases cAMP levels and activity in the receiving cell. There is also a presynaptic norepinephrine transporter, which helps clear norepinephrine from the synapse after neuron firing.

One of the most common signs of depression is a loss of interest in things happening in one's surroundings. Underactivity of norepinephrine-releasing neurons often accompanies depression, and a number of drugs commonly used to treat depression act at least in part by stimulating norepinephrine release from midbrain neurons. Drugs that selectively block the norepinephrine transporter, which increases the concentration of norepinephrine in synapses, are also effective in the treatment of depression. Studies have shown that for many depressed patients, combining SSRIs with drugs that block the norepinephrine transporter are much more effective than either drug given alone. On the other hand, overproduction of norepinephrine may generate feelings of anxiety and fearfulness, as if there were a constant threat present in the environment. Drugs that decrease norepinephrine release are effective in relieving these symptoms.

There is also a possibility that norepinephrine may be a factor in impulsive behavior. Individuals who are unusually sensitive to stimulation by external events may find impulsive responses to them more difficult to control. Indeed, some of the drugs used to control blood pressure by dampening the norepinephrine system turn out to be useful in many cases in controlling impulsive behavior.

Neurotransmitters in Learning and Memory

Learning and memory are among the most important aspects of behavior in humans. They are fundamental to human mental function or "intelligence," which we discuss in detail in Chapter 12. While it might be tempting to think of so-called higher mental functions as somehow special, and different from other behaviors such as aggression, courtship and mating, or impulsivity, recent experiments suggest the very same cellular mechanisms involved in these behaviors are also involved in learning and thinking—they are all mediated by neurotransmitters. Direct experimental evidence for neurotransmitters in human learning and memory is scarce, because we cannot manipulate humans and examine them in the way we do animal subjects. We discuss the rather impressive indirect evidence for a role of neurotransmitters in human mental function in Chapter 12. The most direct and impressive evidence for such a role comes from some recent and rather dramatic experiments in mice, which are worth reviewing here, since almost certainly the results are reflective of what we will eventually find in humans.

Among the neurotransmitters involved in learning and memory is the simple amino acid glutamate. In this case the amino acid is not altered at all; it has been coopted by the nervous system to function as a transmitter in its unmodified form. Glutamate is a strongly activating neurotransmitter. Most nerve cells have receptors for glutamate, and when these receptors are appropriately engaged by glutamate, the cell is pushed into a hyperactivated state. Like most other neurotransmitters, glutamate has a number of different receptors scattered throughout the nervous system, each triggering the cell expressing it to follow a slightly different pathway of activation.

One postsynaptic glutamate receptor, called NMDA, is found in particularly high concentration in the hippocampus region of the brain (Fig. 7.3), where memory is generated. In mammals, the NMDA receptor is involved in long-term potentiation, or the development of long-term explicit memory. But a neuron expressing this receptor will only be hyperactivated, and participate in learning, if two conditions are met: The NMDA receptor must be occupied by glutamate, and simultaneously (or within a matter of seconds) the cell must be activated by a separate signal, coming through some other, different type of neurotransmitter receptor. This is part of the secret of learning to associate two separate events: Two signals must arrive at the cell within seconds of one another. If this happens often enough, the cell is permanently altered in its future responses to these same coupled events.

In exploring this system for learning and memory formation, scientists engineered a mouse that completely lacked the NMDA receptor. As expected, these mice had a very difficult time learning and remembering, especially with respect to spatial relationships. But the story is more complex, and even more interesting. It turns out that the nature of the NMDA receptor itself changes with age; the receptor in older mice is such that it requires that two incoming events occur even closer in time in order for the cell to be hyperactivated. Researchers looking at this system suspected that this could be one reason that older individuals have a harder time learning and remembering than do younger animals. Incoming signals that could be associated by younger animals are being picked up by older animals, but are too far apart in time to trigger a learning and memory response. To test this hypothesis, older mice were genetically engineered to express more of the young-type NMDA receptor in cells of the hippocampus. These mice showed learning and memory (both implicit and explicit) responses equal to those of much younger mice.

The glutamate-NMDA system is present in the human brain, and works exactly as it does in mice. A similar age-related change in NMDA receptor structure receptor occurs in us as well. It may eventually be possible to capitalize on our knowledge of how the mouse system works to improve learning and memory function in humans as they grow older. It is unlikely that this could be achieved through genetic manipulation, as it was in mice, but it may be possible to develop drugs that alter the way the "older" NMDA receptor functions. The NMDA receptor has also been implicated in Huntington's disease in humans, and in schizophrenia. Additional research, largely in animal systems such as the mouse, may shed light on these important human disorders as well.

As the preceding sections make clear, neurotransmitters are intimately involved in many distinctive human behaviors. Is this involvement causative, or simply correlational? In other words, are the higher or lower levels of neurotransmitters seen in various behavioral states a primary cause of the behavioral alteration, or do the altered neurotransmitter levels simply reflect psychological alterations underlying the change in behavior? This is an important question, and not one we can readily answer by studying human behavior alone. As we see in the chapters that follow, there is ample evidence from the study of animals that very similar changes in behavior can actually be induced by experimentally altering neurotransmitter levels, which provides strong evidence that altered neurotransmitter levels are causative, rather than reflective, of altered behavioral states.

Genetic Control on Neurotransmitter Pathways

There are a number of ways in which the genes an individual inherits could influence the functioning of neurotransmitter pathways. One way would be simply in regulating the amounts of various neurotransmitters that neurons have available to release. All neurotransmitters are the end products of enzymatic pathways present in the cells that produce them. Different alleles of the genes governing production of these enzymes could lead to differing rates of neurotransmitter production in different individuals. We already have in hand evidence that the gene for a key enzyme involved in serotonin production, tyrosine hydroxylase, is present in multiple alleles in humans. As we will see in the next chapter, one of these alleles shows a highly significant association with impulsive aggressive behavior in males. Alleles of other genes involved in neurotransmitter production are also known, and researchers are now testing their association both with neurotransmitter levels in various regions of the brain and with inherited behavioral patterns.

Allelic variants of genes could also affect both the number and the sensitivity of neurotransmitter receptors on neurons, leading different individuals to respond differently to the same levels of a given neurotransmitter. Both pre- and postsynaptic receptors would be expected to influence the workings of a given neurological pathway. Again, there is abundant evidence that the genes for many of these receptors are polymorphic in humans. Multiple alleles have been found for the genes encoding the transporter receptors for all of the neurotransmitters discussed in this chapter, and these differing alleles correlate with different levels of neurotransmitters present in the system. One allele of the serotonin transporter gene has been associated with anxiety; another allele has a strong association with depression. An allele of the serotonin-1β regulatory gene has been associated with Tourette's syndrome. Most of the postsynaptic dopamine receptor genes are polymorphic. Specific alleles of the D3 dopamine receptor gene have been associated with novelty seeking, and an allele of the D5 receptor gene is associated with a range of psychiatric disorders. While intriguing, these data are preliminary at this stage, and will require further confirmation.

No one imagines that any one of the alleles of neurotransmitter-related genes uncovered so far explains any single behavior completely. As we have just seen, behavioral patterns such as impulsivity and depression are affected by multiple neurotransmitter pathways. None of the neurotransmitters we have introduced here works in isolation; each affects the pathways of the others, and drugs aimed at one pathway often influence another pathway as well. What is obvious is that the spectrum of genetic variants in just the pathways discussed here can have enormous implications for variations in human personality and behavior.

Imagine a large roomful of people, each with his or her unique collection of alleles for the genes encoding the generating enzymes for these neurotransmitters. Then imagine the possible combinations of these differing transmitter levels with slightly differing personal constellations of the respective receptors: An overzealous serotonin transporter, say, paired in the same individual with an underactive postsynaptic dopamine receptor in a particular region of the brain. Or a poorly functioning norepinephrine regulatory receptor coupled with high levels of dopamine caused by an underactive dopamine transporter, and an overall low level of serotonin. Or any of these working in conjunction with a young versus an old NMDA receptor. Think of all the possible combinations in a room of just a few hundred people. How will those combinations play out in the context of environmental events taking place in that same room? A threatening individual entering the room, background music playing too loudly, a smoke alarm going off, or a waiter dropping a tray of drinks on someone? How will environment interact with genotype to create the highly individual reaction phenotype that is Phyllis, George, or John? What about a pair of identical twins in the same room, with their identical genes for all neurotransmitter elements, both enzymes and receptors? Will they always respond identically to every event in the room? If not, why not ?

Understanding how neurotransmitters and their receptors function has been a major step forward in understanding how the brain works, and especially in understanding how the brain regulates behavior. The evidence that allelic differences in neurotransmitter-related genes can cause behavioral differences is accumulating rapidly; what remains is not to prove this point, but to further dissect how changes in the underlying genes, and their interactions with one another and with the environment, influence behavioral patterns. Beyond doubt, these elements will be key in any final understanding of the genetic basis of behavior. In the next few chapters, we look at several of these behaviors in more detail.

9

The Genetics of Aggression

Aggression, particularly in human beings, is almost as difficult to define as behavior itself. Instructors in psychology classes, at the beginning of the section of lectures about aggression, sometimes ask students to write down their own definitions of aggression. The resulting descriptions are highly idiosyncratic, and invariably as broad as they are vague. Unfortunately, the same question put to an international meeting of professional psychologists would elicit definitions that, while couched in more sophisticated jargon, would probably be only slightly more focused than those tendered by college freshmen. Psychology straddles a number of disciplines, from the social to the biological, with specializations as disparate as criminal behavior, social welfare, and the interplay between the mind and the immune system. Each of these subdisciplines of psychology would approach the subject of aggression from a different point of view, and their practitioners would likely come up with noticeably different definitions.

But mutually acceptable definitions are important if scientists are to communicate with one another. John Renfrew, in his book *Aggression and Its Causes*, offers the following definition: "Aggression is a behavior that is directed by an organism toward a target, resulting in damage."* This is a useful definition in that it identifies aggression firmly as a behavior, and strips it of qualifiers that would make it in any way uniquely human. Aggression is one of the most widespread and fundamental of behaviors, only slightly less ubiquitous among living organisms than reproduction itself. It stems ultimately from the force that has shaped the evolution of all life—competition for limited resources and breeding partners in the drive to reproduce. From both a genetic and an environmental point of view, aggression is a phenotype to be measured and accounted for like any other.

Although rather minimal, Renfrew's definition of aggression has embedded in it several important points. It categorizes aggression as a *directed* behavior; the object of the behavior is generally another individual, of the same or a different species. The purpose of the aggression may be immediately fatal to the object of aggression, for example, when one animal eats another (predation), or in those species where males claiming a new female may kill her dependent offspring sired by other males. However, aggression toward an individual can be indirect, such as in the destruction of another's nest or burrow, in which case the immediate object of aggression may be an inanimate object. Also important is the inclusion of the notion of damage. In the case of predation, the damage to the individual object of aggression is obvious and total. In other cases, the ultimate damage is simply to render the individual less able to compete for the resources needed for reproduction. This may involve physical harm to the individual, but it doesn't have to, as in the case of destruction of nests or eggs.

There are genuine behavioral differences between male and female mammals, and this is certainly true of aggressive behavior as well. But it is true of only certain kinds of aggression, namely, the type of competitive aggression occurring between members of the same species, the biological basis for which is competition for breeding partners or establishment of hierarchies within social groups for preferential reproductive rights. This is the form of aggression we examine in this chap-

---

* John W. Renfrew, *Aggression and Its Causes: A Biopsychosocial Approach* (New York: Oxford University Press, 1997), p. 5.

ter; among nearly all mammalian species, it is a type of aggression engaged in mostly by males. In general, females are relatively nonaggressive in competition relating to reproduction, although there are a few exceptions among higher primates. Female baboons, as we have seen, engage in overtly aggressive behavior in establishing preferences for breeding with the dominant male of their troop. But in general, females generally show low levels of this type of aggression unless they are pregnant or nursing.

Males fight for control of territory, to ensure an adequate supply of food, and they must defend this territory once established. They also fight among themselves for females to breed with, and must continually ward off male intruders with the same design. Studies have shown repeatedly that male reproductive success is directly correlated with aggressive behavior in virtually all mammalian species. From this biological perspective, aggression is a highly positive, albeit largely male-oriented, behavioral trait. Although substantial physical harm may occur as a result of aggression, it is rarely fatal. When two males fight, dominance of one over the other is usually established rather quickly. Most species have clear submissive postures or other mechanisms by which a male can signal acceptance of defeat. This results in a rapid cessation of attacks by the dominant male. It is also usually accompanied by chronic stress patterns in the submissive male in future encounters with the dominant male. Blood levels of stress hormones are always elevated in submissive males when the dominant male can be seen, or even when his presence is detected through pheromones.

Human behavior introduces additional variables into the study of aggression. For example, language makes possible the practice of verbal aggression. Animals often use threatening verbal noise and gestures as an accompaniment to aggressive acts, but in humans the aggression may lie entirely in words themselves, for example, in written threats. In humans, and possibly some of the higher primates, we also must make a distinction between impulsive versus premeditated aggression. It is in the former that a role for genes is most directly evident. Some behavioral psychologists have labeled all subhuman aggression as impulsive, but it is not entirely clear that we mean the same thing when we talk about impulsiveness in humans and animals. We do not really know what "action without sufficient reflection" means in regard to mice, for example. On the other hand, although language, cognition, and culture make all human behaviors more complicated to dissect, there is no reason to believe that the basic *causes* underlying human aggression are different from those we see in other animals, or that genes play a greater or lesser role in regulating aggressive behavior in humans than in other animals.

Genetic Contributions to Aggressive Behavior

The fact that differences in male and female aggressive behavior are so uniform across so many different mammalian species makes it highly likely that this difference has a genetic basis. Since gender is itself genetically determined, it is difficult to escape the conclusion that many of the behavioral differences in males and females are also affected by genes. But additionally, we would like to know to what extent *variability* in aggression seen within the same sex in a given species also has a genetic basis. In other words, to what extent are the genes involved in behavioral differences between the sexes also involved in differences within the same sex? We thus begin our discussion with a brief consideration of the genetics of gender differences in mammals.

Gender is determined by the type of sex chromosomes an individual possesses: Females have two X chromosomes (i.e., they are XX), while males have an X and a Y chromosome (they are XY). The smaller Y chromosome derived from the X chromosome during evolution by repeated gene loss, and the Y chromosome today has very few genes compared with most chromosomes. All of the Y chromosome genes appear to be involved either with the generation of sperm or with the determination of male gender. When genes on the X and Y chromosomes are compared carefully, it can be seen that most of the Y chromosome genes evolved from genes on the X chromosome, although corresponding genes on the two chromosomes have diverged from one another to a considerable extent over evolutionary time, and no longer control the same phenotypes–that is, they are no longer alleles, but have become distinct genes.

The critical point for expression of Y chromosome genes in determining maleness in mammals is during embryonic development. The earliest stages of this development are referred to collectively as the "sexually indifferent stage," because development of the genital tract up to that point is the same in both XX and XY embryos. At some point, the genes on the Y chromosome begin to be expressed, and the path of development of the genitalia after that point in Y-bearing individuals is male. The development of female genitalia is in a sense the default state in mammals: If expression of certain key Y chromosome genes is suppressed or disrupted, the individual will turn out to be female, even though genotypically XY. This is reflected in certain rare genetically altered states in humans. In Turner's syndrome, individuals are born who are X0—they have only a single X chromosome inherited from their mother, but no Y chromosome contributed by the father. These individuals have female genitalia, showing that a single X chromosome is sufficient for development of femaleness. In Klinefelter's syndrome, infants are born with two (or occasionally more) X chromosomes in addition to a Y chromosome–for example, they may be XXY, or even XXXY. Although these individuals show increasing degrees of feminization, they still develop male genitalia, showing that two or even more X chromosomes cannot overcome the tendency of the Y chromosome to direct primary sexual development toward maleness.*

The key to Y chromosome control of gender is formation of the testes during embryonic development. Once the testes are formed, they begin producing two key hormones, testosterone and the so-called anti-Müllerian hormone. Testosterone promotes the further development of male sexual structures, while the anti-Müllerian hormone shuts down further formation of female structures. The gene on the Y chromosome largely responsible for this development has been isolated, and given the name *sry* (sex-determining region of the Y chromosome). It is found near the tip of the short (p) arm of the Y

---

* Although the external genitalia are normal in these two conditions, development of ovaries and testes is usually compromised, causing sterility and impairing formation of secondary sexual characteristics such as body shape, breasts, vocalization, and facial hair. There is also a higher than normal incidence of mental retardation, although some individuals with these syndromes are intellectually unimpaired. Some males are born with an extra Y chromosome (XYY). In one study, it was found that the incidence of XYY males is significantly greater in prison populations, but whether this is related to excessive aggressive behavior, or to other deficits associated with an extra Y chromosome (e.g., learning and reading difficulties) has not been established.

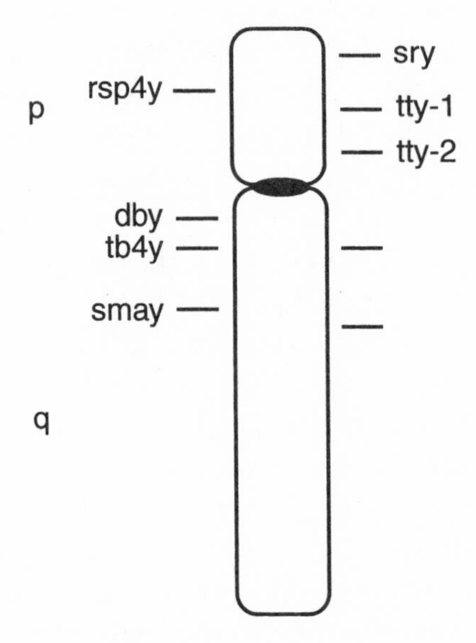

Figure 9.1 The human Y chromosome. Like all chromosomes, the Y chromosome consists of s short (p) and a long (q) arm, connected by a dense centromere. Genes unique to the Y chromosome, expressed mostly in the testes, are listed on the right side of the figure; those shared with the X chromosome, and expressed throughout the body, are shown on the left. Most of the far q-end of the Y chromosome contains no genes at all.

chromosome (Fig. 9.1). The *sry* gene is one of those rare genes in which there is no variability in the entire human species. When the *sry* is compared among closely related species such as humans, chimps, and gorillas, it is ten times more different than other genes.

The *sry* gene was identified by studying rare cases of sex-reversed humans who are genetically XY, but phenotypically female. The Y chromosome in such cases is not quite normal, however. Many such individuals had lost the same small portion of the Y chromosome, and this was inferred to be the locus of a male-determining gene. The *sry* gene was subsequently found at this site on normal Y chromosomes, and it was further found that in those cases where sex-reversed individuals had not lost the section of the Y chromosome containing *sry*, the *sry* itself gene was mutated. Final proof of the male-determining function of *sry* was obtained in mice, where an *sry* gene alone, introduced into XX embryos, resulted in the production of XX male individuals. The Sry protein encoded by the *sry* gene looks like those proteins that bind to and regulate the expression of genes, and the best guess at present is that Sry initiates the expression of genes involved in the development of male genital structures, particularly the testes.

In addition to its association with gender, which indicates a genetic basis, degrees of aggressive behavior are heritable, which also strongly suggests the involvement of genes. Heritability of aggressiveness can readily be shown in mice, where it has been possible to breed selectively for high- and low-aggression mouse lines. One way of testing aggression is to place two males, who have had no previous contact, in a single cage with a clear partition between them. After several minutes, during which they become aware of each other's presence, and perhaps analyze visual cues, the partition is raised to allow physical interaction, and chemical interaction through exchange of pheromones. The mice are then scored in terms of how long it takes them to begin to interact physically, and the direction and strength of aggression displayed during their ensuing interactions. This can range from simply smelling the other mouse, to bumping, kicking, wrestling, and biting.

The most and least aggressive mice are selected for mating, and their offspring are then tested in the same way. Those offspring are again sorted on the basis of high or low aggressive behavior, mated, and their offspring tested and selected in the same way. Within only a few generations, the high- and low-aggression lines show clear separation (Fig. 9.2), with gradually smaller changes after half a dozen or so generations. Aggression in males of both lines could be lowered to essentially zero by castration prior to puberty.

At first it seemed that aggression was being bred only into males in these studies. Females tested in a variety of pregnancy–nursing situations showed no difference in aggression between high- and low-aggression lines. Injection of testosterone into adult females of each line had little effect. But when females of the aggressive line were injected with testosterone at birth and then again as adults, they became as aggressive as males of the aggressive line. Similar treatment of females of the low-aggression line had no effect. This shows that while presence of a Y chromosome per se is not absolutely required for aggressive behavior, it is very likely required for development of testosterone expression. Moreover, there must be genes on autosomes that are important in mediating the effect of testosterone on aggression, and females must express these genes as well as males.

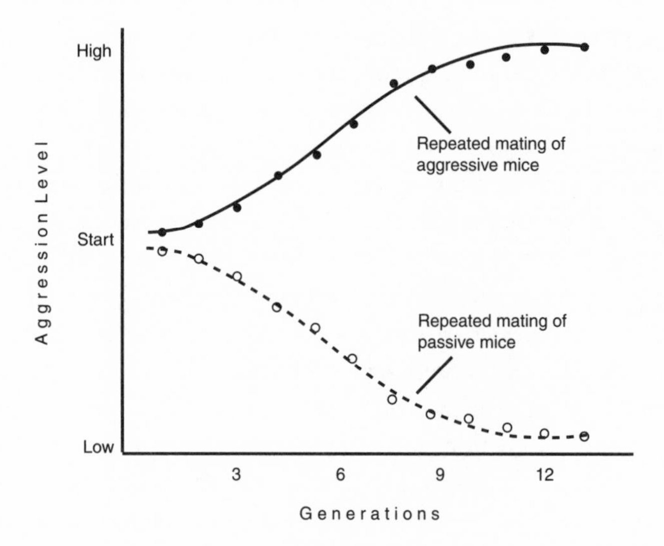

Figure 9.2 Selective breeding for aggression in mice.

Aggressive behavior breeds true within a given mouse strain, and is not learned from adult members of the strain. Placing newborn high-aggression pups with a low-aggression foster mother for nursing, and subsequent rearing with low-aggression cage-mates, has no effect on their aggressive behavior as adults. The same is true for nonaggressive pups reared by an aggressive mother and aggressive cage-mates; they are still nonaggressive as adults. This provides strong evidence for a significant genetic component of aggression in these animals.

That aggressive behavior is heritable in humans is suggested by studies with monozygotic and dizygotic twins, reared together or apart. In a study of twins from the Vietnam Era twin registry, data from over 300 fraternal and identical male twin pairs were analyzed after testing for four aggression-related behaviors. Each of these behaviors had a significant heritable component: direct (physical) assault, with a heritability of 47 percent; verbal assault, 28 percent; indirect assault (temper tantrums; malicious gossip), 40 percent; and irritability, shown to be highly correlated with aggressive behavior, 37 percent. All four of these aggression subtypes are considered impulsive. A significant effect of nonshared (but not of shared) environmental influence on variability in aggressive behavior was detected, ranging from 53 percent for direct assault to 72 percent for verbal assault.

All this implies a significant genetic influence on aggression, and thus an underlying chemical basis for its operation within an individual. We can guess from Figure 9.2 that aggression will almost certainly be a quantitative trait, and that more than one gene will likely be necessary to explain such a complex behavior. But what might the genes for aggressiveness encode? We don't yet have all the answers in hand, but some candidates are now well established.

The Chemical Basis of Aggression

Hormones. The best known chemical mediator of aggression is the male sex hormone, testosterone. Testosterone is primarily involved in inducing aggression among males competing for social position and mating preference, and to a lesser extent in interspecific aggressive behaviors such as killing animals for food. Testosterone, though synthesized primarily in the testes and to a lesser extent in the adrenal glands in males, is also present at low levels in females, where small amounts are made in the ovaries and in the adrenal glands. Recent evidence suggests that testosterone may be important, along with a form of estrogen called estradiol, in the aggression displayed by females while pregnant or nursing. It has also been implicated in human female aggression. However, neither estrogen nor any of its related metabolites by themselves play any role in female aggression.

Testosterone is not a direct product of a gene; it is not a protein. Rather, it belongs to a family of chemical structures known as steroids. Testosterone is known as an "anabolic" steroid, meaning among other things that it promotes tissue growth. In this role it is responsible for the extra muscle mass in males. It is also responsible for cosmetic differences in males, such as a deeper voice, more body hair, and in some males less hair on the head. It is also a factor in higher cholesterol levels in men, contributing to the greater incidence in males of cardiovascular disease. Like the principal female sex hormone, estrogen, and other steroid hormones, testosterone is actually synthesized from cholesterol, and its synthesis depends on numerous enzyme-controlled reactions. Testosterone is closely related structurally to estrogen. In fact, testosterone is metabolized in males to produce a testosterone derivative called dihydroxytestosterone (DHT), and a form of estrogen called estradiol. Both of these forms may play a role in male aggressive behavior, as we see below. The cellular receptor for testosterone is itself a protein. Unlike most receptors that cells use to receive extracellular signals, which are embedded in the cell membrane pointing outward, steroid receptors are found floating free in the cytoplasm of cells. Testosterone diffuses passively across the cell membrane, binds to its receptor in the cytoplasm, and then enters the nucleus of the cell where the complex of testosterone and its receptor activates genes sensitive to the presence of this complex. Thus, while testosterone will enter virtually any cell in the body, it will activate genes only in those cells that have a cytoplasmic receptor.

That testosterone is a major factor underlying aggression in male rodents has been made clear by numerous studies with both rats and mice. In both unselected rats, and in rats selectively bred for aggressive behavior, male aggression is virtually eliminated by castration prior to puberty, and it can be fully restored by administration of testosterone. A role for testosterone in mediating aggressive behavior in human males is generally recognized, but interpretation has been less straightforward. We cannot (and would not want to) do the kinds of experiments in humans we do in animals, and we must rely on measured correlations between circulating levels of testosterone and observable aggressive behavior. This runs us into sampling problems: Testosterone levels are very definitely under circadian rhythm control, and can fluctuate by a factor of two or more depending on the time of day the sample is drawn. We also do not know how things like mood affect testosterone at different points in an individual's day or week. While in general human males classified as aggressive have higher levels of testosterone, this is not always the case.

One possible explanation of different responses to the same level of testosterone in different men could lie in the gene encoding the testosterone receptor, or the receptors for its metabolites. Genes for the testosterone receptor are known to be polymorphic in humans, and these allelic differences appear to affect the ability of these receptors to bind to testosterone. Two individuals with different receptor alleles would respond quite differently to the same level of testosterone. Moreover, there are differences in the number of receptors expressed per cell in different individuals. We do not know how receptor number is determined, but such differences are known to exist, and would also affect an individual's response to a given testosterone level, and possibly the duration of the response.

The brain contains numerous cells with testosterone receptors, particularly in the hypothalamus, which is a site of synthesis of many other hormones. Consistent with the lack of effect of estrogen on aggression in females, there are no estrogen receptors in these same regions of the brain where testosterone receptors are found. The precise circuitry of aggression has not been worked out in the brain; electrical stimulation of some areas increases aggression, while stimulation of other areas suppresses aggression. Nevertheless, the role of testosterone in aggression is amply documented, and both production of and sensitivity to this hormone is considered to be central to an understanding of aggressive behavior.

There have been attempts in the past to discount testosterone as a factor in human male aggressiveness, but such attempts are simply not consistent with the evidence. It is highly unlikely that the correlations of testosterone with aggressiveness observed in other mammals do not apply to humans. The lack of simple correlations of circulating testosterone levels and aggressive behavior in humans has numerous explanations consistent with what we know of how testosterone works: multiple aggression-initiating pathways in the mammalian brain, each sensitive to different testosterone metabolites; allelic differences in the receptors for testosterone and its metabolites; and differences in the number and cellular distribution of testosterone receptors in the body. All of these qualities have an underlying genetic basis, and variations in the genes involved would certainly be expected to influence variability in human aggression.

Neurotransmitters. A second major category of molecules in the body that has been associated with aggressive behavior is neurotransmitters. Although a number of neurotransmitters have been implicated in aggression, particularly in humans, by far the most intensely studied is serotonin. In mice, a role for serotonin in modulating aggression was suggested by the fact that, among genetically inbred mouse strains, male aggressive behavior is inversely correlated with serotonin levels in the brain. That is, the most aggressive mouse strains have the lowest level of serotonin in the brain and its associated fluids, while less aggressive strains have higher serotonin levels.

The role of serotonin in aggressive behavior in animals became particularly evident with the development of a group of drugs called "serenics," which greatly reduce aggressive behavior in male mice, making them more "serene." Once scientists had isolated and studied the structure of the various serotonin receptors, it became possible to design serotonin "look-alike" drugs—molecules that look very much like serotonin, but which interact only with particular serotonin receptors, such as those found on neurons. The serenic drugs have this property. When administered to mice, they bind primarily to the serotonin transporter. This receptor is located at the tips of the axons of serotonin-releasing neurons (see Fig. 8.1). When the transporter is blocked by a serenic drug, the local (synaptic) serotonin concentration increases, thus strengthening the transmission of serotonin-mediated signals. On the other hand, these drugs have no effect on other potentially serotonin-sensitive cells expressing different serotonin receptors. It was found that serenics reduce aggressive behavior in mice without triggering any of the body's other responses to serotonin.

The role of serotonin in mouse aggression was made absolutely unambiguous by the production of "knockout mice" for yet another serotonin receptor. Knockout mice, so-called because a single gene has been deleted, or "knocked out" from their genomes, have become a valuable research tool for studying the effects of individual genes. It is possible to isolate the very early cells from a developing embryo, and disrupt within them a single, specific gene. These embryo cells, lacking a functional copy of the gene in question, can be reintroduced into a developing embryo in the uterus of a pregnant female mouse, and by appropriate breeding of the resulting offspring, a strain of mice can be derived that is completely lacking just a single gene. These kinds of mice are exactly like mice in which a gene has been lost or disabled through a natural mutation.

This technique has recently been used to produce mice completely lacking in another serotonin receptor, called the 5HT1β receptor. (5HT is scientific shorthand for serotonin, which is formally known as 5-hydroxytryptamine. 1β simply refers to the particular subtype of serotonin receptor.) The 5HT1β receptor senses the level of serotonin in the immediate area of the cell, and regulates the cell's ability to release serotonin. When the 5HT1β receptor is completely knocked out, the serotonin releasing pathway shuts down altogether, and the level of serotonin falls precipitously in those pathways using neurons dependent on the 5HT1β receptor. "Downstream" neurons that would ordinarily be activated by serotonin lie completely inert. The knock-out is engineered to affect serotonin pathways only in the brain; other serotonin-sensitive cells throughout the body do not perceive anything to be wrong. 5HT1β receptor knockout mice display much more intense aggression than control mice that are identical except for having the 5HT1β receptor gene. Knockout males attack a strange male mouse much more quickly, and more viciously, than controls. Pregnant or nursing knockout females also show increased aggressive behavior.

The great value of the knockout mouse is that it makes clear that decreased serotonin function can be a *cause* of aggression, and not merely a side effect. If decreased brain serotonin were caused by some mechanism operating only in aggressive mice, that is, were only a *result* of aggressive behavior, then lowering brain serotonin levels in normal mice should have no effect on aggression. The fact that serotonin-impaired mice are always highly aggressive is important in interpreting the role of serotonin in aggressive behaviors generally.

The role of serotonin in human personality disorders involving aggression has been under intense study for several decades. Decreased concentrations of serotonin have been detected in a wide range of individuals displaying aggressive behavior, for example, in children showing repeated outbursts of aggressive behavior at home and at school, in young men discharged from the military services for repeated violent behavioral episodes, in criminals incarcerated for repeated crimes of violence, and in many others. The type of aggressive behavior associated with low serotonin is impulsive, rather than premeditated, aggression. As we have seen, a major function of serotonin in humans is to regulate impulsive behavior. Individuals often know full well that the consequences of certain behaviors may be negative for them, but they have difficulty controlling these behaviors if their serotonin concentrations fall below a certain level.

A major criticism of the studies relating to serotonin and aggression in humans is that they have been carried out largely on psychopathically disturbed individuals, and thus the results of these studies may tell us nothing about aggression among nondisturbed individuals. In a sense, this is a circular argument, since excessive aggressive behavior was in many cases one of the traits leading to the classification of these individuals as disturbed in the first place.

There has also been a reluctance to apply information gained about serotonin function in animals to interpretation of human behavior. A major difficulty in translating the results of the correlates of aggression in animals to humans is the lack of simple, clear definitions and measures of aggression in humans. Humans probably do possess counterparts to aggressive behaviors like predation and reproduction-related aggression in males, and pregnancy/nursing-related aggression in females. But these behaviors are mingled with, and partially camouflaged by, other biologically and culturally dictated behavioral overlays, and these are not always easy to quantitate. Many older studies included anger in measurements of human aggression, and this now seems to be a separate behavior entirely. Analyses of aggressive behavior in humans often rely on self-reported activities, or the impressions of family members and friends, as well as testing and observation by various trained professionals. These methods are all indirect, but we cannot simply put humans in cages, manipulate their hormones or neurotransmitters, and then put them through defined behavioral tests. Finally, as noted earlier, there is some reluctance to equate impulsive behavior in humans with aggressive behavior in mice, even though the role of at least serotonin in mice and humans appears to be indistinguishable.

The use of serenic-type drugs to modulate human aggressive behavior generally confirms a role for serotonin in human as well as animal aggression. Several drugs are available that block the serotonin transporter in humans, effectively increasing its general concentration within synapses along serotonin pathways in the brain. Chief among these are the SSRIs discussed in Chapter 8, such as Prozac, Paxil, and Zoloft, primarily used for the treatment of depression and mood disorders. In a recent clinical trial involving forty patients, the effect of Prozac on impulsive/aggressive behavior was assessed. The individuals involved had histories of such behavior, but were not suffering from major psychiatric disorders such as schizophrenia, manic-depressive disorder, or major depression, and they were not substance abusers. After ten weeks of taking the drug, there was a substantial decrease in impulsive/aggressive behavior among those taking the drug compared with those receiving placebos. A number of individuals dropped out of the study along the way—a major difference between human and animal trials!—and several individuals did not respond to the drug. However, the drug dosages used were not particularly high, and future trials will both enroll larger numbers of subjects and look at the effect of increased drug dosages for nonresponders.

For obvious ethical reasons, it has not been possible to correlate aggression and serotonin in humans using many of the strategies successful in animals, such as gene knockouts. However, a measurable degree of serotonin reduction has been achieved in both animals and humans through dietary restriction of the amino acid tryptophan, from which serotonin is produced. Serotonin reduction through this means does not cause aggression per se, but in both animals and humans, responses to subsequent provocations in the environment are measurably more aggressive.

Recent experiments with another knockout mouse support the notion of deregulated neurotransmitter function in aggressive behavior. Serotonin, norepinephrine, and dopamine are all cleared from synapses by presynaptic transporter receptors. Once reabsorbed into the releasing neuron, they are degraded by an enzyme called monoamine oxidase A (MAOA). When the *maoa* gene was disrupted in mice, brain concentrations of all three transmitters increased significantly. Male *maoa* knockout mice were found to be much more aggressive than their normal counterparts. It is difficult to know exactly which neurotransmitter caused the increase in aggression, but increasing either dopamine or norepinephrine in animals had previously been shown to increase aggression. Elevated norepinephrine in humans has also been found to correlate with increased aggression. It is also possible that hyperexpression of serotonin may lead to increased aggression, since aggressive behavior in the knockout mice was attenuated somewhat by serotonin inhibitors. Variabilities in aggressive behaviors could thus be regulated at least in part by varying levels of serotonin and other neurotransmitters in different individuals.

The *maoa* gene in humans, which is on the X chromosome, has multiple alleles; although we do not fully understand how the various gene products differ functionally, it is entirely conceivable that they will have varying degrees of efficiency for degrading resorbed serotonin. Recently, it has been shown in a study of several generations of a Dutch family that unusually aggressive behavior in nine male family members is very closely linked to the *maoa* gene. Affected individuals were very quick to anger in situations they peceive as threatening, and anger turned to violence in a very high proportion of instances. Violence took the form of assault, rape, arson, and attempted homicide and suicide, among other things. All nine individuals showed varying degrees of mental retardation as well. A subsequent study of the *maoa* gene itself in these individuals showed that a slight internal mutation had resulted in production of a protein that was completely functionless. While the genetic components of aggression are almost certainly complex, involving numerous possible genes, this study makes clear that changes in a single gene can have a major effect on a behavioral phenotype.

Serotonin is made in the body from the amino acid tryptophan, and a key enzyme controlling its synthesis is tryptophan hydoxylase. It was recently shown that there are two allelic variants of the gene for this enzyme in humans. One allele, U, is present in 40 percent of the population; the other allele, L, is present in 60 percent. It was found that males who were homozygous for the L genotype (L/L) were significantly more impulsively aggressive than heterozygous individuals (U/L), or individuals homozygous for the U allele (U/U). However, this was not true of females; L/L females were no more aggressive than U/U females, suggesting the effect may be testosterone-dependent. Finally, there are also different allelic forms of most of the genes encoding serotonin receptors expressed in the brain. Although we do not yet know the functional implications of the allelic differences we see in these genes, it is entirely possible that these different receptor forms will be shown to have differing affinities for serotonin.

Aggressive behavior has recently been correlated with another neurotransmitter, nitric oxide. The recognition of nitric oxide as a neurotransmitter is itself a fairly recent development. Nitric oxide is a gas, and the notion that a gas could be involved in communication between nerve cells was difficult to accept at first.* Moreover, this is a highly

---

*The 1998 Nobel prize in medicine or physiology was awarded to three scientists who pursued another role for nitric oxide, namely, its ability to cause blood vessels to relax and dilate. The common heart medicine nitroglycerin, for example, works by releasing nitric oxide, which dilates blood vessels, including those serving the heart. The nitric oxide system is also the target for another popular drug, Viagra.

toxic gas; macrophages, scavenger cells of the body that engulf and destroy marauding bacteria, use nitric oxide in the destructive process. It is a major component of automobile smog, and is thought to be a factor in destruction of the ozone layer. But neurons do, in fact, produce nitric oxide and release it into their immediate environment. Unlike all other neurotransmitters, which bind to and enter only cells displaying a specific receptor, nitric oxide simply diffuses across the membrane of any cell in the vicinity, which, in the brain, is usually another neuron. Once taken up by another neuron, nitric oxide binds to iron molecules attached to one of the key enzymes involved in signal transduction within nerve cells, thus producing a signal just like any other transmitter.

Nitric oxide is produced in neurons by an enzyme called nitric oxide synthase (NOS). The gene encoding this enzyme in brain cells is different from the enzyme carrying out the same function in other tissues. In a recent study, scientists took advantage of this to make knockout mice specifically lacking the neuronal form of NOS (nNOS). The nNOS knockout mice appeared normal, displaying no ill effects in physiological systems using other NOS enzymes to generate nitric oxide—the knockout was indeed specific for nNOS. But the effects on behavior in NOS knockout mice under appropriate test conditions turned out to be quite marked. Like the serotonin receptor knockout mice, male nNOS knockout mice were incredibly aggressive, attacking strange males more quickly and more vigorously than controls with the nNOS gene intact. They also failed to respond to normal signals of submission by victims of their attacks. Normally, when an attacked male rolls over on his back and extends his limbs straight outward, the attacking male will desist, and further fighting is minimal. Knockout males continued to attack males who had clearly signaled surrender.

In addition to excessive aggression against male intruders, nNOS knockout males also showed heightened sexual activity toward females. A normal male, when presented with a female in estrus, will mount her within minutes, and repeat this activity for about fifteen minutes, after which the male gradually loses interest. The knockout males continued trying to mount females many hours after normal males would have ceased, in spite of vigorous vocal and physical protests from the female. Since both aggression and sexual activity are known to be affected by testosterone levels in males, the level of this hormone was tested in the knockout mice. The levels were not different from normal males. They also performed the same as control mice in an open-field test.

So far, only male nNOS mice appear to be more aggressive, but nursing nNOS females have so far not been specifically tested. In fact, they become less aggressive, not even defending their young when threatened. Castrated nNOS knockout males do not show increased aggression, but do so if they receive injections of testosterone. This shows that the loss of NOS, and thus nitric oxide as a neurotransmitter, activates a testosterone-dependent pathway of aggression in nNOS male mice. Since knockout males do not show elevated testosterone levels, it is possible that nNOS loss may affect some other aspect of the testosterone pathway, for example, the number or distribution of testosterone receptors, or signaling pathways within neurons affected by testosterone.

Just how altered levels of neurotransmitters affect aggressive behavioral patterns is not yet clear. We do know that in rodents neurotransmitter and hormonal components of aggression are interrelated. Both male and female mice that are injected over several weeks with excess testosterone show markedly decreased brain serotonin. Castration of male mice before puberty results in elevated brain serotonin compared with controls. In the case of nitric oxide, it is clear that increases in aggression caused by deletion of the NOS gene require testosterone, since castrated NOS knockout males do not show increased aggression. Unfortunately, castration experiments have not yet been done on serotonin receptor knockout mice, so it is difficult to assess the role of testosterone in the case of serotonin-deficient animals. However, the fact that female serotonin receptor knockout mice also show increased aggressive behavior raises the possibility that serotonin could be affecting a testosterone-independent pathway.

Aggression is not a behavior that occurs in isolation from other behaviors. It contains a strong element of impulsivity, and we can see clearly that drugs affecting other behavioral and emotional overlays, such as depression, have a profound effect on aggressive behavior as well. As with any behavior, it is a propensity to violence and aggression that is inherited, not the behavior itself. All of the measures of aggression that have been tested in humans have a substantial environmental component as well. The heritable irritability component of aggression may suggest a heightened sensitivity to nearly everything in the environment, which would agree with the finding of increased norepinephrine in many aggressive individuals. Events in the environment can trigger aggression, but environmental factors such as education, counseling, and simply life experience may also be important in learning to manage a tendency to aggressive behavior. This enriched understanding of the full scope of human behaviors—encompassing both genetic and environmental factors—has greatly expanded our ability to both diagnose and treat more effectively individuals with a wide range of potential behavioral problems.

10

The Genetics of Consumption, Part I

Eating Disorders

Aside from animals living in captivity, eating disorders are a uniquely human problem. For most of our evolutionary history, the major dietary challenge facing humans was the same as for any other species—simply getting enough to eat, in terms of both the calories needed to fuel our metabolic machinery and the specialized substances like vitamins and minerals required to make that fueling efficient and effective. Proper nutrition is still a major health problem in most underdeveloped and some developing countries. Most dietary health problems in technologically advanced countries, including the United States, stem not from undernutrition, but from overeating: the consumption of calories in considerable excess of our metabolic needs.

The most obvious result of overeating is obesity. Some storage of body fat is not only normal, but absolutely required for good health. Excess body fat, on the other hand, can be extraordinarily dangerous, and nearly a third of all Americans—up from 25 percent just two decades ago—are now classifiable as overweight. These individuals are at risk for a wide range of health problems, including heart disease, high blood pressure, maturity-onset diabetes, arthritis, breast and colon cancer, and many others.

Less immediately obvious, but no less dangerous to human health, are some of the psychologically generated behaviors in response to, or in fear of, obesity. The *Diagnostic and Statistical Manual of Mental Disorders* lists three eating disorders as worthy of the name. Anorexia (anorexia nervosa) results from excessive dieting, sometimes coupled with excessive exercising. It often, but not always, is preceded by a period of overeating leading to obesity. Extreme cases of anorexia result in cachexia, which is the wasted condition of the body we see in starving populations during droughts or other crises where food disappears. It is fatal in as many as 15 percent of voluntary anorectics.

Bulimia (bulimia nervosa) is characterized by repeated overeating (bingeing) followed by purging—the use of vomiting, laxatives, or other means to reverse the consequences of bingeing. Bulemics often suffer from gastrointestinal damage, electrolyte imbalance, and dental damage caused by repeated exposure of teeth to stomach acids in vomiting. Both anorexia and bulemia are often accompanied by depression. "Binge eating" is a recent addition to the manual, and is identified as repeated episodes of overeating, associated with at least three independent behavioral indicators of impulsivity, and not accompanied by purging. Unlike anorexia and bulimia, binge eating affects men and women nearly equally. Interestingly, chronic overeating itself is not presently listed as a disorder, although that may well change in the future; the impulsive nature of some forms of this behavior pattern are becoming increasingly apparent.

Many of the behaviors categorized as eating disorders derive to a large extent from another uniquely human problem: concern about body image, and the perception that our individual "worth" is somehow tied to whether we are fat or thin. While body size and shape are definitely influenced by lifestyle choices, the tendency for thinness and obesity to run in families has been noted for a very long time. This tendency, which would imply a genetic element in obesity, has been confirmed in formal studies of family lineages, and particularly in adoption studies. The degree of obesity of adopted children closely resembles that of their biological parents, and not that of their adopted families. The most compelling evidence for a genetic component of obesity comes from twin studies, in which monozygotic and dizygotic twins, reared together and apart, are compared on the basis of body-mass index, or BMI.* Estimations of the heritability of BMI provide some of the highest intrapair correlations found in twin studies (see Fig. 10.1), and lead to the conclusion that as much as 70 percent of an individual's intrinsic BMI is genetically controlled. Shared environmental factors normally have almost no influence on BMI, with varying degrees of influence of nonshared factors.

The study of heritability of BMI in humans has led to the concept of an intrinsic "set-point" weight, which to a large extent determines the adult body mass for each individual. Nutritionists have long known that chronic overeating or dieting, as well as increased or decreased exercise, can move body weight up or down from a given set point, but as soon as the individual returns to normal eating or exercise patterns, the body moves fairly quickly back to its preset weight. Set-point

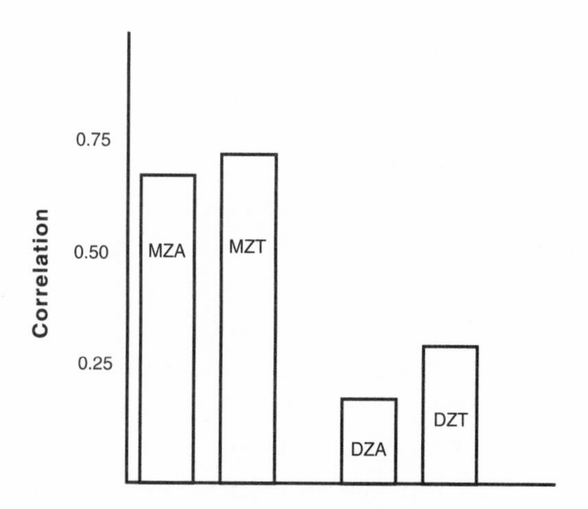

Figure 10.1 BMI correlations in monozygotic and dizygotic twins, reared together or apart. MZA: monozygotic, apart; MZT: monozygotic, together; DZA: dizygotic, apart; DZT: dizygotic, together.

* Body-mass index, or BMI, is expressed as body weight in kilograms (1 kg = 2.2 pounds) divided by the body height in meters (1 meter = 39.4 inches), squared: BMI = kg/m2. A 5'8" person weighing 160 lb would have a BMI of 25.2. A BMI of greater than 24 (for both men and women) is considered overweight; a BMI over 27 is considered obese.

weights may well be above or below a socially desirable "norm," but for the individuals involved, their set-point weight *is* the norm.

In addition to inherited BMI, obesity can also be affected by metabolic differences in the amount of calories individuals with the same BMI require to maintain their set-point weight, and in the readiness with which they convert the food they consume into fat for storage. These factors are controlled by our individual metabolisms, which are definitely under genetic control. But there are two other major factors contributing to obesity in mammals where a genetic connection may in the past have seemed less clear: simple overconsumption of food, and consumption of food with a high fat content.

In general, food preferences are considered to be much more culturally influenced than almost any other aspect of obesity. At first glance, it would seem to be controlled by what is available and by the individual's previous experience with various foods. In terms of preferences for the way specific end-product food items are prepared and presented, this may be true. However, when we look at individual preferences for major food groups—fats, carbohydrates (starches and sugars), and proteins—the available data suggest there may be a significant genetic component at play. The two most important food groups from the point of view of weight control are carbohydrates and fats.

Studies of monozygotic and dizygotic twins indicate a strong genetic component (0.4–0.6 correlation) in preferences for carbohydrates, particularly sweets. In mice, a taste for carbohydrates is definitely heritable. Contrary to popular belief, however, in both animals and humans the strongest preferences for carbohydrates within a heterogeneous weight population are demonstrated by lean, not obese, individuals. For example, when mice are given free access to sugar cubes, leaner mice consume many more calories from sugar than do obese mice. This may be related to a higher rapid-access calorie need on the part of leaner individuals. On the other hand, excess calories taken in the form of carbohydrates will also be stored as fat, leading to obesity.

The genetic preference for fat is somewhat less strong (0.2–0.5 correlation in humans), but in this case the strongest preferences are displayed by obese individuals. In both animals and humans, there is a "diet-induced obesity" that can be traced to diets with a relatively high percentage of fat calories (over 40 percent) in the diet. Two individuals with identical metabolisms, who take in the same total number of calories on a daily basis, will dispose of those calories differently depending on the form in which they are taken in. Calories taken in through fat are more readily stored as fat in body tissues. Thus a genetic predisposition to favor fat in the diet could be a significant factor in "acquired obesity"—obesity above the inherited BMI.

Twin studies, however, also suggest a significant impact of exercise on inherited BMI. Monozygotic twins are essentially genetically identical; differences in body weight between them can be presumed to be due to differences in interaction with the environment. In a recent study that included 241 pairs of monozygotic female twins, in those instances where there was a significant difference in body weight, the single most consistent factor explaining this difference was the level of physical activity. Other environmental factors associated with weight differences were smoking and the use of hormone replacement therapy.

Genetic control of body weight is likely to be highly complex. There will be genes that control the basic body structure plan itself—the set-point BMI. These genes will likely control events in the developing fetus, and may also be at play in adults. Genes that control metabolic rate, or metabolic processes such as fat conversion and storage, will also be important, as will genes that control appetite—the sensation of being hungry or full. Genes also apparently control preferences for particular food subgroups. Any or all of these genes could be at play in the variability we see in obesity phenotypes among humans. Work on sorting out these various genes is just beginning, but already there are some extraordinarily important results providing completely new insights into the genetic regulation of body weight. None is more interesting than the story of leptin.

Leptin: The Body's Own System for Weight Control

Why do we overeat? Is it simply because it is pleasurable? Or may some of us be driven by genetic dictates not under our control? Eating is very much a behavior, and as with any other behavior, how much or how little we eat is ultimately controlled by the brain. Eating behavior is a good example of behavior directed not toward signals generated from the environment, but from within the body itself. But it has in common with environmentally triggered behavior that a signal of some sort must reach a sensory cell connected with the central nervous system—the brain—which will then interpret that signal and set in motion a series of actions to address the problem or need.

Studies in mice over the past two decades provided the first clues to the nature of at least one signaling system involved in telling the brain how much food the body thinks it needs. Key to these studies was the development of several inbred mouse strains in which mice compulsively overeat, becoming vastly overweight and in the process developing many of the same health problems seen in human obesity. Two of these strains, *obese* and *db* (so named because of the high incidence of adult-onset diabetes), ultimately provided the key to understanding the physiological basis of weight control in other mammals. Each strain is characterized by a mutation in a single (but different) gene. The young of both of these strains appear normal at birth, but rapidly commence consuming food as though perpetually hungry. While initial growth is limited by how much food the mother can provide while nursing, they rapidly become excessively obese—up to triple the weight of mice without the underlying mutations—once they are weaned and given unrestricted access to solid food. This trait is passed faithfully from generation to generation.

Several early studies implicated a factor circulating in the bloodstream that controlled the eating behavior in both strains of mice. When an *obese* mouse was joined through its blood vessels with a normal mouse, the *obese* mouse gradually began to eat less and lose weight. This suggested that the *obese* mouse lacked a circulating factor signaling satiety (a sense of fullness), which the normal mouse provided through something in their now-common bloodstream. When a *db* mouse was joined in the same way to a normal mouse, the normal mouse stopped eating and eventually starved to death. The researchers reasoned that the compulsively overeating *db* mouse also produced a circulating factor signaling satiety, but had lost sensitivity to it—perhaps lacking a receptor through which the satiety factor acted. However, the satiety factor would circulate to the normal mouse, convincing it that it was full when it was not, and causing it to stop eating.

In 1994, the gene responsible for the defect in *obese* mice was isolated and found to encode a protein which has come to be known as leptin. Leptin is secreted by fat cells, and is used to signal the brain that the body is consuming excess food that is beginning to be stored as fat. Normally, this results in cessation of eating. *Obese* mice fail to produce a functional leptin protein. The region of the brain receiving this signal is the hypothalamus, which then issues appropriate messages to stop eating. Lesions to the regions of the mouse hypothalamus containing leptin receptors result in compulsive overeaters just like mice of the *db* strain, who were subsequently shown to have a mutant leptin receptor gene. When the altered mice were joined to normal mice, the normal mice also wasted away and died.

The *obese* and *db* strains each demonstrate a different aspect of the leptin pathway for regulation of eating behavior. *Obese* mice have a normal leptin receptor, but fail to produce a functional form of leptin, and cannot signal satiety to their nervous system. Mice of the *db* strain produce normal leptin but, lacking a receptor for it, cannot sense that it is being produced. The result in both cases is hyperphagia, a compulsively eating mouse, a mouse that apparently never feels full. As *db* and *obese* mice accumulate increasing fatty deposits (where leptin is produced), the level of leptin in their bloodstreams rises enormously, but for different reasons they cannot respond to it. On the other hand, when purified leptin is injected into *obese* mice, their weight returns to normal and the physiological derangements accompanying obesity disappear. As expected, leptin has no effect when injected into *db* mice.

The mouse leptin gene was used to search for a similar gene in human DNA, and a corresponding human gene was soon cloned and sequenced. The leptin proteins encoded by the mouse and human genes are 84 percent identical, and both have structural features characteristic of proteins that are secreted into the bloodstream. Yet when blood samples of overweight individuals were screened for leptin, it was found that the leptin molecule itself is normal in the vast majority of these individuals, and that, as expected, the amount of leptin in the bloodstream goes up dramatically with increasing weight. So the explanation for obesity in most humans is not production of a faulty leptin molecule, or failure to secrete it. In rare cases, individuals have been found who, like *ob* mice, are born with defective leptin genes, and these individuals do become severely obese shortly after birth. But this is unlikely to be a general explanation of obesity.

Could it be that humans, like the *db* mouse, have a defect in the leptin receptor? This, too, seems an increasingly unlikely explanation. As with the leptin gene itself, there are a few individuals born with clearly defective leptin receptors, and these individuals also become severely obese shortly after birth. A detailed examination of the leptin receptor gene in large groups of obese individuals has so far not revealed a receptor mutation unique to obese people. But genes do more than just spell out how to make a particular protein. Genes have stretches of DNA associated with them that play no role in the structure of the encoded protein, but which determine where, when, and how the protein is expressed. We know that leptin per se is produced and secreted perfectly normally in most obese individuals, but it is possible that the leptin receptor (which is difficult to measure directly) has errors in its control regions which cause it to be underproduced, expressed in the wrong cells, or expressed in the wrong parts of the right cells.

So researchers are not at all ready to give up on the notion of perturbations in the leptin system as a general cause of human overeating and obesity. They are also working on the possibility that either the leptin molecule may not be crossing the blood-brain barrier to reach its target cells in the hypothalamus in some cases, or that some of the events triggered in brain cells after leptin binds to them are not taking place as they should. We also know that leptin, in both mice and humans, interacts with the insulin system; could there be defects in this interaction in obese humans? Given the resources that are being pumped into this area of research, both by universities and by pharmaceutical companies interested in developing an "antiobesity drug," the site of the defect in the leptin pathway that could explain most human obesity—if it exists—should soon be determined.

There is little doubt that leptin works in humans the same way it works in mice. So even if inherent defects in the leptin system are not the underlying cause of most human obesity, drug companies still hope to be able to take advantage of what we know about leptin to help people maintain a healthier body weight. Leptin may be able to help us move toward a lower body weight without the constant stress of dieting, and the health dangers this can pose, and without an undue lifetime burden of physical exercise. The first step will be to figure out why most obese individuals are not responding properly to their own leptin signals. There may well be additional gene-controlled systems for regulating appetite in humans that either interact with, or are independent of, leptin. Clinical trials with leptin itself have shown that while many individuals are unresponsive to additional leptin, some trial subjects did respond, so it may be that higher levels of leptin could tip the balance in certain subsets of obese individuals.

Drug companies are also working on ways of transporting leptin more efficiently through the bloodstream and facilitating its proper entry into the brain. Proteins as large as leptin do not normally cross the blood–brain barrier unless there is a special receptor in the wall of blood vessels in the brain to facilitate their passage. What is the nature of this receptor, and could it be defective in at least some obese individuals? Still others are examining events inside brain cells once they are triggered by leptin. How do these cells ultimately tell the body to eat, or to stop eating? Now that we know which cells in the brain respond to leptin, we can find out what exactly it is that these cells do, or what they produce—perhaps a hormone that acts on nerve cells or other cells involved in the appetite control pathway. It may be possible to supply this missing product in a suitable drug form.

The development of drugs that can modulate the body's built-in system for weight regulation will almost certainly happen. A combination of health concerns and the desire to look better on the part of obese people will doubtless provide a lucrative market for them. But such drugs will bring their own problems. Leptin-related drugs will probably not change the set point for an individual's inherent weight, so they will have to be taken for a lifetime. We do not know what the consequences will be of long-term consumption of these drugs. And to the extent that these drugs come to displace sensible eating habits and exercise as means of weight control, their overall effect on health may be questionable.

Appetite Control by Serotonin Pathways in the Brain

It has been recognized for over twenty years, largely from studies in rats and mice, that serotonin plays an important role in regulating eating behavior. Eating behavior is controlled by specific regions within the hypothalamus portion of the brain. In general, anything that increases the concentration of serotonin in synaptic pathways leading to these regions reduces food consumption, while anything that decreases serotonin in these pathways increases food consumption.

Chronic interference with normal serotonin release and uptake can lead to excessive overeating and weight gain. When serotonin concentrations or activity are low, the individual feels hungry, and in particular craves carbohydrates. Although carbohydrates do not themselves contain serotonin or its precursor (the amino acid tryptophan), high carbohydrate intake stimulates the metabolic pathways leading to serotonin production. Serotonin levels in the brain rise rather rapidly in response to ingested carbohydrates, and promote a feeling of satiety, curbing further food intake, particularly of carbohydrates themselves. Increased serotonin levels also increase metabolic rate, aiding in the conversion of food to energy. Most animals, including humans, appear to balance the intake of carbohydrates and protein as a source of calories, with calories from protein providing about 12 percent of total intake, and calories from carbohydrates about 60 percent. This natural balancing act is mediated to a large degree by serotonin.

The importance of serotonin pathways in eating behavior was underscored in recent studies using mice in which the postsynaptic serotonin receptor 5HT2c was deleted using the gene knockout strategy described in the last chapter. These receptors are found on neurons leading to regions of the brain known to control eating behavior. The knockout mice were normal in most respects, including structure of the brain and nervous system. But it is interesting that these mice showed a marked tendency to overeat, in some cases coming to weigh nearly twice as much as control mice. The additional body weight was accounted for almost entirely by storage of excess body fat. Using a drug that can suppress appetite by specifically blocking the binding of serotonin to 5HT2c receptors, these researchers showed that whereas such drugs decreased food intake in control mice by 80 percent, they had no effect on the eating habits of the knockout mice. The tendency to overeat was shown to be a behavioral alteration, and not due to changes in how the knockout mice metabolized the food they ate, or their tendency to store it as fat.

The similarity of the effects of leptin and serotonin on obesity raises the question of whether these two pathways may be somehow interconnected. Do defects in the leptin system cause lowered serotonin? Does an inherently low level of brain serotonin interfere with the leptin system? Or are the two defects completely independent? Recent studies with a rat model, similar to the *db* mouse discussed above, provides possible insight into this question. As adults, these rats are grossly overweight, and have very low levels of brain serotonin. Like *db* mice, they also have a defective leptin receptor. Within four days of birth they begin to show signs of fat accumulation and other metabolic changes associated with the onset of obesity. Leptin levels begin to increase in parallel with these changes. However, decreases in brain serotonin do not occur until much later, after the time when gross differences in body weight and fat content are apparent. Thus at least in these rats, lowered serotonin may be secondary to the known defect in the leptin system.

Evidence for a role of serotonin-mediated pathways in human overeating and obesity is based largely on the finding that drugs that interfere with the serotonin pathway in humans cause significant appetite suppression. Fluoxetine (Prozac), for example, blocks the reuptake and degradation of serotonin by releasing neurons, resulting in higher than normal levels of free serotonin in synaptic spaces. This is not a postsynaptic effect, so several serotonin-dependent behaviors, such as aggressivity and depression, are affected in addition to appetite. For this reason fluoxetine is not routinely used for appetite control, but a decrease in eating is a common behavioral side effect of fluoxetine administration. The drug fenfluramine and its derivatives also lead to an elevation of serotonin activity in the brain, and is a powerful appetite suppressant. It is not yet clear whether fenfluramine stimulates release of serotonin, or acts at the receptor level to enhance serotonin signaling pathways. Fenfluramine is part of the "fenphen" drug combination (fenflurazine plus phentermine, an amphetamine-like drug) that was used briefly as an unusually promising appetite suppressant. However, the finding of a high incidence of heart valve problems and pulmonary hypertension in persons taking fenphen led to its withdrawal from the market in late 1997.* Interestingly, recent studies in mice, where the same type of damage is seen, suggest that a combination of fenfluramine and fluoxetine, each in reduced amounts, can achieve the same level of appetite suppression without toxic side effects.

---

* These drugs, sold under trade names such as Pondimin and Redux, were given government approval after very limited testing. They were initially recommended for use only with dangerously obese individuals, but public clamor and doctor willingness to prescribe them led to their widespread use by persons only marginally overweight. The government is now reviewing its procedures for approving these kinds of drugs.

One of the most successful drugs for managing obesity is sibutramine, which is marketed (by prescription only) under the trade names Reductil and Meridia. Like fluoxetine, sibutramine is a serotonin reuptake inhibitor, but it also blocks reuptake of noradrenaline. Sibutramine induces a sense of satiety, at least in the short term, but perhaps of equal importance, it also speeds up the metabolic pathways that burn fat to produce body heat. This increased metabolic activity, probably due to the increase in norepinephrine, may be particularly important in long-term weight reduction. While not everyone responds to sibutramine, those who do are able to maintain a lower body weight for periods of up to two years. The average response to other weight-reducing drugs rarely lasts more than six months.

While there has been some reluctance to define acquired obesity as a form of impulsive behavior, some of the recognized eating disorders are increasingly coming to be recognized as manifestations of impulsivity. The formal descriptions of anorexia and bulimia certainly sound like impulsive behavior—the individuals affected almost always are aware that the behavior they are engaged in is not in their best interest, but they seem unable to break out of the pattern. Support for a connection between eating disorders and impulsive behavior comes from the observation that individuals with eating disorders often have a high incidence of other impulsive behaviors, such as substance abuse or aggressiveness. Conversely, individuals with primary diagnoses of other impulsive behavioral disorders also have a much higher than normal incidence of eating disorders. Family studies are also revealing. Among individuals with eating disorders, there is a two- to three-fold increase in the likelihood of a first-degree family member (a sibling, parent, or child) displaying an impulsive behavior such as alcoholism or drug abuse. The frequency of anorexia in first-order relatives of anorexics is also many times greater than in the population as a whole. Intrapair correlations for monozygotic twins averaged 0.46, and for dizygotic twins, 0.07. The vast majority of studies carried out to date suggest anorexia and bulimia are heritable disorders, with a genetic component of at least 70 percent.

As with overeating, there are strong indications of serotonin deregulation in both anorexia and bulimia. Several studies have found consistently low levels of serotonin metabolites in the cerebrospinal fluid of bulimic women. Many bulimics are depressed, and it is possible that reduced serotonin levels secondary to depression could be responsible for the bulimic behavior. However, antidepressant drugs that increase serotonin levels are effective in treating bulimia whether or not the patient is clinically depressed, suggesting that decreased serotonin levels are likely to be a cause, rather than a result, of bulimia.

A role for serotonin is also generally recognized in anorexia nervosa. Both fluoxetine and fenfluramine are effective in treating anorexia, and as in bulimia, the effectiveness of these treatments seems to be independent of the effects of these drugs on depression. Of particular interest in the case of anorexia is the finding that a genetic alteration in the human 5HT2a receptor for serotonin shows a very high degree of correlation with anorexia. The expression of this receptor is known to be under control of the hormone estrogen, which could provide a basis for the fact that virtually all cases of anorexia involve women.

Serotonin is not the only neurotransmitter playing a role in eating behavior, although it shows the greatest degree of association with eating disturbances. Dopamine knockout mice show a disinterest in eating from birth; when given precursors of dopamine through the bloodstream, they immediately commence eating. In humans, drugs such as amphetamines that block the neurotransmitter norepinephrine also curb appetite. Neuropeptide Y, one of the small protein neurotransmitters, is also involved in appetite control. The amounts of all of these neurotransmitters that are present in the nervous system are under genetic control. These amounts will be different in different individuals because of slightly different allelic forms, scattered throughout the population, of key enzymes involved in their synthesis. Moreover, in almost every case that has been looked at, there are also multiple allelic forms of the receptors for these neurotransmitters within the human population. These differences certainly provide a possible basis for contributions of neurotransmitters to the genetic control of eating behavior.

The Interaction of Genes with the Environment: The Pima Indians and Obesity

In searching for a biological explanation of obesity, we must remember that most of the obesity we see in industrialized countries today came into existence largely in this century, and mostly in the last few decades. It is impossible to explain this sudden appearance of obesity on the basis of a change in our genes during this same period. Genes regulate inherited BMI values, and fundamental genetic changes leading to a change in the distribution of BMI values within a population would take many, many generations—hundreds, and probably thousands, of years. The same reasoning also makes changes in genes involved in the leptin and serotonin pathways an unlikely primary explanation of the recent changes in obesity.

We can imagine that there would have been strong selective pressure against inherited obesity when our ancestors were largely hunter-gatherers. The need to move quickly over long distances, plus the requirement for large amounts of food to maintain a higher BMI, would have made obesity a genuine liability. Obesity-favoring alleles of leptin or its receptors, or of genes that control serotonin pathways in the brain, would also have been selected against. Certainly among the few remaining hunter-gatherer societies that have been studied in modern times, obesity is very rare. We can also imagine that when agricultural societies came into existence some 10,000 or so years ago, the selective pressure against higher BMIs, or unfavorable leptin alleles, may have eased somewhat, and that over time some level of genetically determined obesity may have become biologically tolerable. But this almost certainly has been a minor contribution to contemporary obesity.

We definitely do see a limited amount of genetically determined obesity in our society today. But the great changes we have seen in the incidence of obesity in this century are changes in acquired, not inherited, obesity. These changes result from voluntary overeating and underexercising; they are a behavioral problem. As such, they certainly can have a genetic basis, but as with the genes controlling heritable forms of obesity, the genes controlling behavior are unlikely to have changed in the past 100 years or so. So why do we have so many more obese people today?

The answer must certainly lie in changes in the environment, and can largely be accounted for by three things: a greater availability of food generally; a decreased dependence on natural foods, together with an increased availability of high-fat "junk" foods; and a decreased dependence on physical exercise in the accomplishment of life's basic tasks, such as simply getting from one place to another. Our collective genotypes have not changed significantly in the past century, but the environment in which these genotypes must operate has changed dramatically, and this can bring about an equally dramatic change in the *phenotypes* we see around us. Depending on our individual genotypes, we will have responded to these changes in different ways. Depending on our genetically determined behavioral propensities, we may or may not engage in behaviors that result in obesity; we may or may not respond to our fear of obesity by engaging in anorectic or bulimic behavior. So while the recent changes in the incidence of obesity have been triggered largely by changes in the environment and not by genetic changes, our individual responses to those changes—the phenotypes we express in response to those changes—are in the end largely genetically determined.

The interplay between genes and the environment in the development of obesity is spectacularly evident in the recent history of the Pima Indians. The Pima have lived in the Gila and Salt River basins of Arizona for at least the past 2000 years. Throughout most of this period they subsisted by irrigation farming, supplemented by hunting and fishing. This way of life was disrupted at the end of the 1800s, however, when European settlers began moving into the Arizona desert in large numbers, and diverting water from the rivers used by the Pima for irrigation. This disruption was somewhat ameliorated by completion of the Coolidge dam project in the late 1920s, which provided both the Pima and the descendants of the European settlers with adequate water. However, the Pima, like other farmers in the area, used the new water source largely to produce crops for sale on the open market, as a means of generating cash, rather than for subsistence. Moreover, a few decades later, water demands in nearby Phoenix dried the Gila River completely, bringing farming and fishing essentially to a halt. The Pima now consume processed foods as a major part of their diet, and have adopted a much more sedentary lifestyle.

The Pima lived in great poverty until fairly recently. As part of a hardscrabble desert farming culture, they were far from the American mainstream and generally ignored. They were granted U.S. citizenship only in 1924, and were granted the right to vote in Arizona only in 1948. Males were, however, drafted for service in World War II, and others were recruited to war production jobs far away from the reservation. The war saw an approximate doubling of per capita income among the Pima. Nevertheless, the return of soldiers from the war, and the gradual drift back to the reservation of wartime workers from various urban centers, marked yet another major cultural disruption for traditional Pima.

The Pima had been studied extensively as part of a project gathering data on the physical attributes of native Americans in 1905, and were not found at that time to be particularly obese. But after the Second World War, the Pima began showing signs of unusual obesity. This obesity begins quite early in life; children are already markedly overweight well before they begin school. By age thirty or so a large proportion of both men and women have BMI values of thirty-five or even more. The average five-foot, ten-inch male weighs about 240 pounds; women are even heavier. This obesity is accompanied by all of the normal morbidities associated with excess weight. For example, the Pima have the highest known incidence of adult-onset diabetes of any identifiable group in the world: Over half of adults suffer from this disease.

The environmental causes of the recently acquired obesity among the Pima are clear: a change in diet from low-fat, naturally grown fruits and vegetables, supplemented by occasional meat, toward a diet mainly dependent on processed, higher-fat foods, plus a marked decrease in physical exercise. Estimates of the fat in Pima diets at the end of the last century range around 15 percent; by the 1950s this had risen to about 40 percent. But not all Pima become grossly overweight. First of all, there is a related tribe of Pima in Mexico who still live a simpler, agriculture-based existence, and who show little evidence of obesity, underscoring the largely environmental nature of obesity among the Arizona Pima. But of equal interest, there is considerable variability among the Arizona Pima themselves with respect to obesity, and researchers have now shown that this variability is heritable within Pima family lineages. We can thus speculate that there are common human genotypes which, when acted upon by the current dietary and exercise components of American culture, tend toward obesity, and other genotypes that do not. What we see with the Pima is a disproportional slant toward obesity-producing genotypes in the present cultural context. The question now becomes, what are the individual genetic components of the overall obesity phenotype?

To date, the genetic components of the Pima obesity phenotype have not been identified. The leptin system, as far as it is understood in humans at present, appears to be working normally, and no unusual alleles of leptin or its receptor have been found among the Pima. As might be expected, obese Pimas have very high levels of leptin circulating in their systems. But leptin levels fluctuate normally during periods of fasting and eating in the Pima, indicating that the fat storage cells are sending out the correct signals. Studies on neurotransmitters have not yet been carried out, but nothing that is known so far suggests there are major differences in these components either.

Geneticists are hoping that clues may come from the parallel incidence of type II diabetes* among the Pima, which is also the highest in the world. Type II diabetes has sometimes been referred to as a calorically "thrifty" disease, because it encourages the conversion of excess calories to storage forms of fat, which could then be drawn upon over an extended period of food scarcity. It may be that in populations such as the Pima, who historically faced periodic bouts of starvation, the ability to store food as fat in times of plenty would have had positive survival value. In times of persistent high caloric availability, however, the genotype that proved valuable in leaner times may now be a liability. This is of course pure speculation, but it has some intriguing implications. Sorting out the genes involved in type II diabetes is now a very high research priority, in part because these genes may also be involved in the general obesity epidemic plaguing the Western world.

The Pima are an ideal group for analysis of obesity-related genes using the technique of genome scanning introduced in Chapter 5. Several generations of directly related Pima often live in close proximity, making family inheritance patterns easy to follow. In a recently initiated project to scan Pima DNA for obesity-related alleles, the inheritance of various traits associated with obesity were correlated with the

* Diabetes mellitus is divided into two types. Type I, often called insulin-dependent or juvenile diabetes, is caused by a failure to produce a functional insulin molecule. Individuals with type I diabetes are absolutely dependent on an external source of insulin for survival. Type II diabetes, sometimes called insulin-independent or adult-onset diabetes, is a much broader phenotype, involving either suboptimal production of insulin, failure to release insulin from the insulin-producing cells of the pancreas, or failure of insulin to bind properly to insulin receptors on target cells. Treatments for type II diabetes, which accounts for 90-95 percent of all cases of diabetes mellitus, are correspondingly broad and complex.

inheritance patterns of 500 DNA markers scattered throughout the human genome. The traits followed through several generations included percent body fat, as well as two metabolic parameters associated with obesity: energy expenditure levels and the tendency to burn carbohydrates versus fats for energy.

Several genomic regions of interest were detected in this initial study (Fig. 10.2). Each panel of this figure represents a different chromosome, indicated by the number in the upper right portion of the panel. The plots shown here indicate the likelihood (LOD score) that the trait of interest is affected by a gene at a given location on the chromosome. The markers used to study each chromosome are indicated below each chromosome. A representation of the chromosome is laid along the X-axis in each case. (The quantitative trait loci (QTL) on chromosome 11 were all found on the right half of the chromosome; the markers have been moved into the figure for purposes of clarity.)

A very strong association was detected between a QTL on chromosome 2 near marker S1360, and the ability to control the amount of leptin released in response to fat storage. A QTL affecting the tendency to accumulate high levels of body fat was closely associated with marker S2366 on chromosome 11. Also found on chromosome 11 was a separate QTL (possibly two loci) affecting energy expenditure (near markers S976 and S912). Possible QTLs regulating carbohydrate versus fat utilization were found on chromosomes 1 and 20. Intriguingly, the gene for the leptin receptor is close to the chromosome 1 QTL detected in the Pima genome scan. Although the leptin receptor itself is apparently normal in Pimas, as noted earlier it is possible that control regions of the gene may be altered in such a way that the receptor is underproduced in cells, or does not show up at locations in the cells where it is needed.

Researchers are now busy analyzing the regions of the chromosomes shown just below each plot, looking for genes that might be responsible for the associated traits. The next step will be to examine each of these regions with more closely spaced markers. When a small enough region containing the QTL has been defined, scientists will turn to information provided by the Human Genome Project, which by the year 2003 or so will be able to furnish scientists with DNA sequences of the entire human genome. The identity and function of some of the genes found in the region of the QTL will already be known. In other

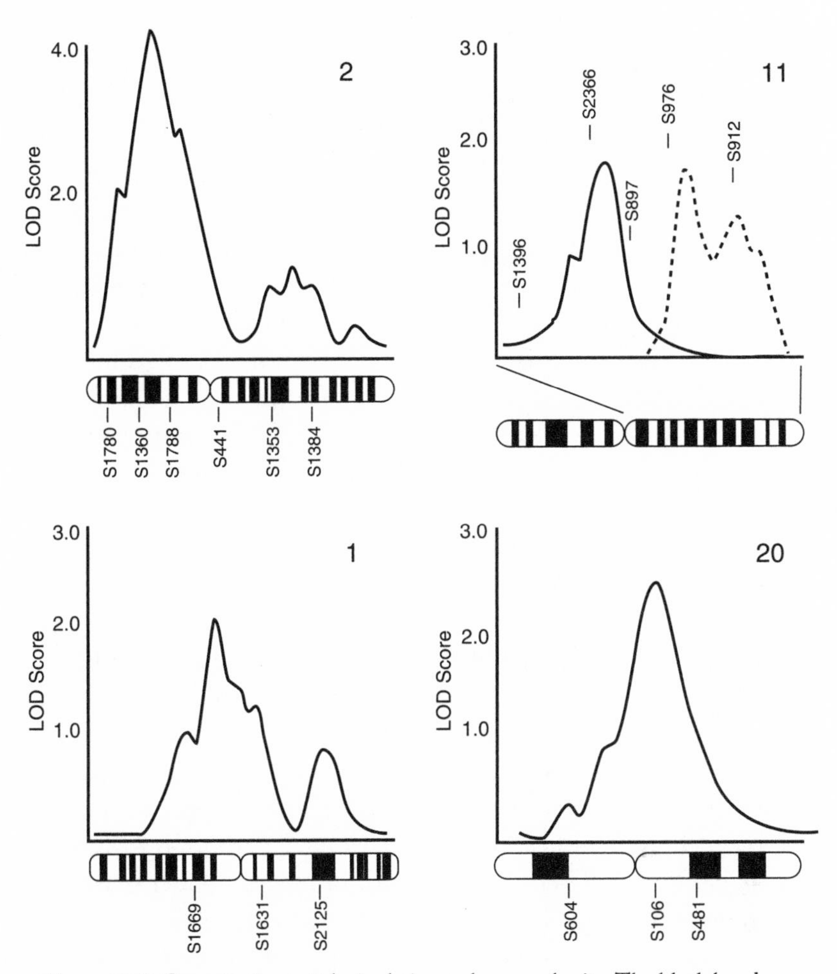

Figure 10.2 Quantitative trait loci relating to human obesity. The black bands are caused by selective dye uptake at certain regions of the chromosome. Chromosome numbers represented are shown in upper right of each panel. Based on data presented in Comuzzie et al. (1997) and Norma et al. (1998). LOD stands for "log10 of the odds"; the higher the LOD score, the more likely the association is real. Only LOD scores of 2.0 or greater are considered significant.

cases it may be necessary to locate the corresponding gene in an animal model, where its identity and mode of action can be more readily determined. This same approach is now being used with a wide range of behavioral traits. (Further details of how genome scans are used to locate genes of interest are given in Appendix I.)

What Lies Ahead?

For most of this century, overeating was treated primarily as a psychological problem. But it is now generally agreed that the rates of definable psychopathologies are not significantly greater, nor is the spectrum of disorders noticeably different, when obese individuals are compared with the general population. More recent psychoanalytic theories of obesity have looked to "unconscious drives" as the basis of compulsive eating, the aim of which is seen as reduction of anxiety or stress. This description is not far off from what we now think of as impulsive behavior, a connection made more attractive by the demonstrated involvement of serotonin and other neurotransmitters in eating behavior, and the strong association between eating disorders and other forms of impulsive behavior.

It is indeed possible that some behavioral phenotypes, such as impulsive behavior, may make it more difficult for certain individuals to resist environmental factors that contribute to obesity in modern society. This has not been well studied, and it probably should be. But if this is true with regard to diet and exercise, such individuals will be helped not by books promoting yet another miracle diet, or some strange new confabulation of foods and herbs and vitamins and creams, but by sound advice and monitoring by a health professional well versed in *both* the genetic and behavioral underpinnings of weight control in humans. There is no scientific or medical evidence whatsoever to support diet plans that emphasize one food group over another. With the general rule of keeping fat calories to no more than one-third of total calories consumed, the most effective diet consists of no more than reducing total caloric intake in the context of a nutritionally balanced diet. To be most effective, any dieting should be accompanied by a moderate level of exercise.

The information we have in hand at present tells us that how one goes about losing weight depends very much on the origin of the weight problem. If one is among that minority of people who are overweight because of an inherited high BMI, dieting and exercise may be beneficial in shedding unwanted pounds, but keeping them off will be a lifelong battle because the body will resist weight reduction with everything at its disposal. If one has a lower BMI, but is obese because of overeating combined with a sedentary lifestyle, the message is quite clear. It is indeed possible to get back to one's normal BMI. The difference in this case is that there is a very good chance that any weight lost will stay lost, as long one does not engage in the voluntary abuses that led to obesity in the first place.

While sensible eating habits and exercise programs will always be an important component of any weight control program, the leptin story suggests that the future may provide other means to help us control excess body weight. Leptin has been an exciting discovery, because we now know that the genetic components of something as complex as eating behavior can indeed be dissected out. The failure so far to find a direct connection between obesity in most humans, and alterations in leptin or its receptor, simply makes the challenge more interesting. Scientists have by no means finished their exploration of the leptin system itself. But they are already asking whether humans may have other genetically controlled pathways regulating food consumption, food storage, and appetite. If so, what are these pathways? Do they interact with leptin, or are they completely independent? Those are exactly the kinds of challenges scientists love.

The discovery of the leptin system could well be just the first step toward uncovering genes whose variabilities within a population may explain the variability we see in eating behavior and obesity. Genome scans will almost certainly lead to the discovery of additional genes important in producing the obesity phenotype. Once these various components are identified, and we understand how they work and how they interact with environmental factors, it is not at all unreasonable to expect that we will be able to develop meaningful strategies for weight control that bypass dieting and exercise regimens that are simply unrealistic for some individuals. Molecular medicine is already making inroads into a wide range of maladies affecting the human body; weight control, with all of its implications for human health, may well be next.

11

The Genetics of Consumption, Part II

Substance Abuse

Overconsumption of food carries serious health risks for the obese individual, and is a burden to society to the extent that those risks eventually require publicly supported health care for the individuals affected. But obesity does not even begin to compare with disorders such as alcoholism and drug abuse, which can be absolutely devastating both to the individuals involved and to their families, and a major strain on society as a whole. The key element that distinguishes eating disorders from alcohol or drug abuse is addiction. Addiction is a form of repetitive behavior from which an individual is either unwilling or unable to refrain, in spite of harmful consequences. This sounds like something akin to impulsive behavior, and indeed there may well be an element of impulsiveness in the beginning stages of addiction. But in true addiction, there is an additional dimension of physical or psychological dependence—or both—that makes control of the underlying behavior exceedingly difficult. What may have begun as impulsive behavior soon becomes compulsive in nature.

As with virtually all other forms of behavior, there is a substantial contribution of genetics to the tendency to abuse drugs. This has been shown by a great many family and adoption studies over the years, and most convincingly by a study of over 3000 twin pairs as part of the Vietnam Twin Studies Registry. Substance abuse has also been studied in monozygotic and dizygotic twin pairs, reared together or apart. These various studies estimate the genetic contribution to addictive behavior as somewhere between 30 and 50 percent. There is, for example, an eight-fold increase in the likelihood that a first-degree relative of a substance abuser will have a problem with drug addiction, compared with the general population. Most studies to date have suggested that the heritable tendency toward drug abuse is not substance-specific; the tendency is toward substance abuse in general. There is also a strong correlation of drug addiction with major depression. Depression, too, has a strong genetic component, but the degree of overlap in genes involved in substance abuse and depression is not entirely clear. In many cases, individuals can be cured of one problem independently of the other, suggesting that the overlap is certainly not complete.

Heritability patterns for substance abuse correlate strongly with heritability of other personality factors. Among individuals seemingly predisposed to substance abuse there is a higher than normal frequency of individuals manifesting impulsive behaviors, such as attention deficit disorders, thrill seeking, aggressiveness, and gambling, among others. There is also a strong correlation of substance abuse with negative emotionality, including personality parameters such as neuroticism, anxiety, and alienation.

What sorts of genes might we expect to influence the behavioral parameters of substance abuse? To the extent that impulsivity is a contributing factor, we might expect an involvement of certain neurotransmitters. But is that the whole story? To appreciate the possible genetic bases of substance abuse, we really need to know as much as possible about the cellular and molecular components underlying the physiology of addiction.

For some drugs, particularly hallucinogenic substances, addiction may be primarily psychological or emotional. For other drugs, such as alcohol or some of the opiates (e.g., heroin), there is a definite element of physical dependence as well. In either case, genes appear to be a contributing factor, but the underlying mechanisms may be quite distinct.

Physical dependence is judged by the body's reaction to withdrawal of a particular drug. Withdrawal from "hard-core" drugs like heroin may involve fever, nausea, pain, diarrhea, and loss of appetite. Surprisingly, withdrawal from seemingly milder drugs like barbiturates and alcohol can be even more severe, leading, in some cases, to coma and even death. The basis for physical addiction can be traced back to the effects of drugs on individual cells. Many drugs that induce physical dependence interfere with the way individual cells ordinarily function. Under continual use, the cells adapt to the presence of the drug by altering their internal metabolic pathways, many of which involve cAMP as a signal-transducing agent. All is well as long as the drug, which is part of the newly altered system, remains in place. But when the drug is withdrawn, the cells cannot immediately revert back to the original pathway. For a period of time, they continue to operate via the altered pathways, a key component of which is now missing. The result is that during the transition period these pathways do not operate very well, and the cumulative effects of large numbers of dysfunctional cells underlie the painful symptoms associated with drug withdrawal.

Addiction has other components as well. The phenomenon of tolerance occurs when repeated use of a substance induces a state wherein more and more of the substance is needed to achieve the same effect. Tolerance is induced rapidly and profoundly with some drugs, such as heroin and cocaine, and to a lesser extent with drugs such as alcohol, tobacco, and barbiturates. The normal clinical dose for the use of morphine in controlling pain is about ten milligrams for an adult human, but someone abusing morphine may eventually require a thousand or more milligrams for pain relief. As we see below, tolerance, too, can be understood at the cellular and molecular level.

In terms of what they have in common, drugs in each of these categories can be thought of as positive reinforcers, at least in the early stages of addiction. This does not mean they reinforce positive behavior; it simply means that taking them often enough, across a short enough period of time, increases the likelihood that they will be taken again. But once established, addiction is almost always accompanied by development of a negative physical or emotional state, in which the drug becomes a negative reinforcer as well—the drug becomes necessary to avoid the unpleasant physical or emotional side-effects of drug withdrawal. The resulting addictive behavioral patterns often displace behavior that is more favorable to the individual's well-being, such as eating well and getting sufficient rest. Why is this so? How do these drugs gain such a powerful hold on the neurological mechanisms underlying behavior? And to what extent can the variability in human susceptibility to addictive drugs be explained by genetically determined variabilities in these mechanisms?

Although we are far from understanding the complete physiological basis of addiction, it is clear that, as with all other behaviors, neurotransmitters and their receptors again play a key role. And while the range of potentially addictive substances is enormous, in fact the neurological mechanisms underlying addiction are remarkably similar for all of them. A key element tying all addictive substances together is their ability to coopt the body's primary system for rewarding behavior—the dopamine pathways of the brain, described in Chapter 8. In the sections that follow, we take a closer look at what is currently known about the cellular and molecular bases for addiction to three categories of commonly abused drugs: the opiates (morphine and heroin), cocaine, and alcohol. We also look at the evidence that genetics plays a role in the predilection to addictive behavior, and possible sites in the pathways leading to addictive behavior where genetics might operate.

The Opiates: Addiction to Morphine and Heroin

The opiates are a class of substances obtained from the milky fluid found in the seed pods of the opium poppy. Opiates have been used throughout recorded history as a source of analgesics, or pain killers. The use of opium in a nontherapeutic sense, for its general euphoria-inducing properties, is less well recorded, but certainly goes back hundreds, if not thousands, of years. The most important naturally occurring analgesic compounds in opium are morphine and codeine. Morphine was first isolated in pure form at the beginning of the nineteenth century, and proved to be a potent analgesic. However, its highly addictive nature has, since the very beginning, limited its clinical usefulness to all but the most serious pain-producing situations. In addition to being an analgesic, morphine is a general physiological depressant, slowing activity in the circulatory, respiratory, and digestive systems. This is itself of considerable value when treating someone who has been seriously injured; it calms the individual down at the same time it dulls the pain. However, this, too, places limitations on its use; life-threatening collapse of these systems can result from excessive doses of morphine.

Codeine, the other natural component of opium that is useful medically, is a moderate and frequently used analgesic. It is also an effective cough suppressant, acting directly on the cough center of the brain. Codeine is considerably less addictive than morphine; occasionally people who use it on a long-term basis for cough suppression may experience mild cravings in its absence, and they may find that they need to increase the dosage over time to maintain effectiveness in controlling their coughs. Serious addiction problems, however, are rare.

There are also numerous chemically synthesized variants of opiates that are used clinically, designed to maintain analgesic properties without the complications of addiction. Fentanyl is one such drug; it is a highly effective analgesic, but it is cleared from the body so quickly there is less opportunity to set in motion the sequence of events leading to addiction.

Heroin, the most commonly abused opiate in contemporary society, is also not a naturally occurring compound. It was first prepared as a chemical derivative of morphine at the end of the nineteenth century. It is a much more potent analgesic than its parent compound, but also more addictive. Unlike morphine, its clinical use was greatly curtailed shortly after it was first introduced, and in fact its manufacture is now illegal in most countries of the world. Like morphine, heroin induces a transient state of euphoria in its users, and reduces feelings of fear and anxiety, which contributes to the psychological aspects of addiction to both drugs. Both heroin and morphine are also classic examples of the phenomenon of tolerance in drug addiction. Within a fairly short time after initial use, the amounts required to achieve the same effect, whether analgesic or euphoric, greatly increase. This has further limited the clinical value of opiates, and of course is a major factor in the sociology of opiate addiction.

The analgesic effects of opiates are mediated through the body's natural system for interpreting and managing pain, and a full understanding of how this system operates is vital to an understanding of opiate addiction. Pain is an important part of the biology of all multicellular organisms. It alerts an animal to the fact that something is wrong, and initiates behavioral maneuvers that address the source of the pain. Responses to pain are often reflexive, such as withdrawing a hand from a flame. Sometimes an appropriate response to pain requires more complex behavior: climbing a tree to get away from an attacking animal, or removing a thorn from under a nail or claw. And sometimes there is little or nothing that can be done to modify the source of pain. Kidney stones are enormously painful, but until quite recently in human history the pain simply had to be endured. But pain is rarely ignored.

There are two distinct aspects to pain, and each can be separately manipulated or modulated. The first element is simply the transmission of pain from either an internal or an external pain receptor to the brain. For the most part, these receptors, located on sensory neurons, detect heat, touch, and various chemical stimuli. In humans, pain signals travel to the brain through the spinal cord. The intensity of pain transmission is something the brain is able to control by mechanisms we explore below. The second element of pain is pain perception. Like any other signal coming into the body through sensory neurons, the brain must integrate and interpret pain signals, and make a decision about what they represent.* Decreases in either the transmission or the perception of pain are referred to as analgesia.

The key elements in the body's management of pain are the peptide neurotransmitters β-endorphin, the enkephalins, and dynorphin. The existence of these compounds was detected only in the 1970s, as a result of research aimed at understanding the body's response to opiates derived from or related to morphine—hence the term "opioid" (opiate-like) for these neurotransmitters. All three play a role in both the transmission and the perception of pain. They can act either in the brain or along the spinal cord to reduce the intensity of an incoming pain signal, through the activation of interneurons that suppress the pain-detecting sensory neurons. They also interact with the brain dopamine system described earlier to alter the perception of pain, as we see below.

* While all multicellular animals formulate behavioral responses to the transmission of what we call pain, it is not obvious at what point in evolution the *perception* of pain as discussed here first appears—probably somewhere during evolution of the vertebrates, but we cannot be certain.

As with most other neurotransmitters, these neuropeptide transmitters interact with multiple receptors scattered throughout the nervous system. The three major receptors are called μ (mu), δ (delta), and κ (kappa). The presence of these receptors on dopamine-releasing neurons in the mesolimbic portion of the brain, which is involved in the sense of reward relating to particular behaviors, is particularly important in both pain perception and in drug addiction. β-endorphin and the enkephalins interact with both μ and δ receptors to increase dopamine release of neurons, while dynorphin interacts with the κ receptor to decrease dopamine release. These receptors were first identified on the basis of their interaction with opiate drugs, not with neurotransmitters. The μ receptor, for example, was so named because of its ability to bind morphine long before the existence of β-endorphin and the enkephalins was even suspected. The ability of opiates and opioids to bind to the same receptors is surprising, because they are not at all alike structurally. The opiates are complex organic molecules belonging to a chemical class called alkaloids, which are a group of slightly alkaline compounds obtained mostly from plants, whereas the opioids are all small proteins. Nevertheless, it is quite clear that opioids can block access of opiates to a given receptor, and vice-versa.

It seems highly likely that stimulating dopamine release in the brain, which brings a sense of well-being, is also at play in altering the perception of pain. Wounded animals under attack use their opioids to set aside concern about pain in order to focus on escape. Opioids are also thought to be responsible for the sudden passiveness of certain animals in the extremes of agony, for example as they are being eaten by a predator while still alive. Humans suffering from massive, irreversible physical trauma, for example major combat wounds, often behave in a similar fashion in their final moments of consciousness and life. Humans also use endogenous opioids to modulate pain perception during extreme athletic activities that push the body to its limits of endurance. Whether opioids also induce significant levels of euphoria is less clear, although the sense of relief and well-being that trained athletes feel after vigorous exercise may be a mild expression of this effect.

Opiate drugs act as analgesics in two ways, reflecting the two different aspects of pain. First, opiates interfere with the transfer of signals from the sensory neurons that detect pain to the interneurons in the spinal cord that carry these signals to the brain. Second, the opiates alter the perception of pain in the brain. This is a subtle and sophisticated distinction. Some modulated pain signals still reach the brain even in the presence of opiates, but the interpretation of these signals as representing something painful is altered. Persons treated with morphine for pain often comment that the pain is still there, and they are perfectly aware of it, but they simply don't care about it as much.

The cooption of endogenous opioid pathways in the addictive effects of opiate drugs can readily be demonstrated in animals. Rats become addicted to opiates just as humans do, and as far as we can tell all of the elements of addiction are similar in both species. Drugs can be delivered either intravenously or directly into the brain, and after a short period of trial and error rats will self-administer opiate drugs by pressing a lever. Rats will also become addicted to their own opioid peptides if these are administered in concentrated form; the lever-pressing activity they engage in to self-deliver opioids is indistinguishable from their behavior in obtaining opiates. They demonstrate both tolerance, in the form of constantly increasing numbers of lever pushes required to achieve a given effect, and dependence—severe malaise when the drugs are withdrawn—to both opiate drugs and their own opioids. Administration of substances that block the opioid receptors on their neurons greatly decreases the ability of opiates to induce addiction. Moreover, mice with the μ-receptor gene knocked out are not only insensitive to morphine as an analgesic, but do not easily become addicted to injected morphine.

Opiate drugs affect neurons in several different regions of the brain in the sequence of reactions leading to addiction. Euphoric effects are caused by the action of opiates on dopamine-containing neurons in the brainstem, particularly the mesolimbic region. Dopamine released by these neurons acts on other neurons in pleasure and reward centers of the brain such as the nucleus accumbens, causing positive sensations of well-being. Opiates can also act directly on the neurons in the nucleus accumbens, so it is clear that stimulation of these neurons is a vital element in the reward system. Dopamine release from mesolimbic neurons under the influence of potent opiates such as heroin is extraordinarily high, and occurs very rapidly, producing the intense "rush" described by addicts.

Studies in animals and on nerve cells isolated in culture have provided insights into how opiates alter the function of neurons. The major change induced by opiates is a marked increase in cAMP-mediated activities in neurons of the mesolimbus. The increased cAMP results in expression of new genes in these cells, which is absolutely essential for the euphoric effects of these drugs, and also in regulating their overall effects. Excess exposure to drugs may induce the synthesis of dynorphin, which counteracts their effects and helps keep the brain functioning within reasonable bounds. Opiates also act directly on neurons of the nucleus accumbens to cause an increase in cAMP.

Some of the new genes induced by cAMP are thought to be involved in countering the perturbing effects of opiates on cellular metabolism. This in turn may be involved in the development of physical dependence; when opiates are withdrawn after a period of continuous use, it takes several days for the cAMP system in these cells to return to normal. The cAMP circuitry involved in drug dependence is the same circuitry involved in learning and memory. Rats in which the cAMP-induced expression of new genes was shut down were more resistant to the effects of opiates, but they also suffered major deficits in memory formation.

Cocaine

Cocaine is an alkaloid extracted from leaves of the coca plant, which have been used for at least 1500 years by peoples indigenous to South and Central America as an analgesic and to some extent as a stimulant. Taken in larger amounts or purer form cocaine, like the various opiates, produces a state of euphoria. But it is a euphoria accompanied by heightened mental awareness, exhilaration, relief from fatigue, and increased physical activity, as opposed to the generally relaxed and dreamlike state that accompanies opiate-induced euphoria. That may well have been a factor in its inclusion (until 1906) in small amounts in the original formula for the soft drink Coca-Cola. By 1914, cocaine was placed on the list of narcotic drugs in the United States. Today it is one of the most widely abused substances in the United States: A 1994 survey carried out by the U.S. General Accounting Office found that nearly a million people had used cocaine *at least once a week* during the preceding year.

Cocaine is among the most positive of reinforcers among currently abused drugs, and tolerance to the drug builds very rapidly. Cocaine is also cleared from the system more rapidly than most other drugs; together with the tolerance effect, this can lead addicts to rapidly achieve levels of usage sufficient to kill if administered to a drug-free individual. Yet paradoxically cocaine does not create nearly the same level of physical dependence as the opiates (judged by withdrawal symptoms), suggesting that addiction is more closely tied to psychological dependency. That this sort of dependency can be highly potent is indicated by the extremely low success rate for rehabilitating cocaine addicts, and there is currently no effective treatment for cocaine addiction. For reasons that are not entirely clear, the concordance between cocaine use and alcohol abuse is higher than for almost any other drug.

As with opiates, dopamine pathways in the mesolimbic portion of the brain play a major role in the psychotropic effects of cocaine. Shortly after cocaine enters the bloodstream, there is a rapid rise in dopamine concentrations in pleasure centers of the brain, such as the nucleus accumbens. The same rises in intracellular cAMP triggered by opiates are seen in response to cocaine, particularly in the nucleus accumbens. Interference with the dopamine pathway blunts cocaine addiction. But cocaine does not act through opioid receptors to activate these pathways. Rather, cocaine blocks presynaptic dopamine reuptake receptors in the mesolimbus, decreasing the clearance of dopamine from synapses and thus leading to an increase in dopamine in the brain.*

However, addiction to cocaine may not be quite as exclusively tied to the dopamine system as opiate addiction seems to be. A knockout mouse strain in which the gene for the dopamine transporter has been disabled shows, as expected, a greatly reduced sensitivity to cocaine. Dopamine levels are already very high in the brains of these mice, and the self-administration of cocaine does little to increase these levels. Nevertheless, with patience these mice can be induced to become addicted to cocaine, suggesting that with this drug there may be pathways other than those involving dopamine.

---

* Interestingly, a dopamine reuptake receptor sensitive to cocaine was recently discovered in *C. elegans*, showing how evolutionarily ancient this system is.

Identifying these alternate pathways has become a top priority for those struggling to understand the cellular and molecular basis of cocaine addiction. Since cocaine inhibits the serotonin transporter even more strongly than the dopamine transporter, and also interacts with the norepinephrine transporter, repeated cocaine usage would be expected to raise brain levels of these neurotransmitters as well. However, drugs that selectively block the serotonin and norepinephrine transporters do not themselves cause either euphoria or addiction. Experiments with mice lacking the serotonin transporter gene, which increases brain serotonin, and mice lacking the 5HT1b receptor, which decreases it, both affect cocaine addiction, but not in ways that clarify exactly how serotonin is involved. Sorting out the contributions of these neurotransmitter pathways to cocaine addiction may ultimately provide insights into potential treatments for cocaine addiction in humans.

Although cocaine does not bind to opioid receptors, there is some evidence that it may nonetheless mobilize the opioid system during the course of addiction. Drugs that block opioid pathways in the brain greatly interfere with establishment and maintenance of cocaine addiction in animals. Several times more cocaine is required to induce addictive behavior in rats when cocaine is administered with the drugs naloxone or naltrexone, which block μ opioid receptors. Recent evidence indicates that opioids may also be involved in the phenomenon of tolerance to cocaine. Dopamine released by cocaine-stimulated neurons in the nucleus accumbens induces the synthesis, via the cAMP pathway, of dynorphin. This opioid acts through the κ receptor to offset the reward effects of cocaine. In rats, chemicals that stimulate κ receptors (thus mimicking dynorphin) are highly effective in preventing cocaine addiction.

Genetically different inbred strains of mice have widely differing natural propensities to drug addiction, suggesting that differing alleles of key genes may control various aspects of the addiction process. Advantage is being taken of this to look for quantitative trait loci that influence drug addiction, on the assumption that identification of such QTL in mice, and subsequent identification of relevant genes at these loci, could greatly accelerate our understanding of addiction in humans. The work of sorting out candidate genes located at these QTL sites is just beginning, but already apparent are genes encoding various dopamine receptors, as well as several hormone receptors. In one sense the involvement of dopamine receptors comes as no surprise, but QTL studies have the possibility of making it absolutely clear that inheritance of different alleles of a given gene are associated with different behavior in response to a particular drug. And of course, most intriguing of all, QTL studies have the potential of identifying previously unsuspected genes involved in a given behavior.

Finally, some recent and intriguing results in Drosophilia show how complex the biology of cocaine addiction may be. Flies can also become addicted to cocaine, and will essentially drop out of all other normal activities in order to pursue sources of cocaine. But flies with mutations in their *per* or *clock* genes, as well as other genes involved with their circadian rhythms (*tim* is an exception), are highly resistant to cocaine addiction. The molecular basis for this is of course of great interest in terms of understanding human addiction, but it also increases the kinds and numbers of genes in which variability may be related to variations in human addictive behavior.

Alcoholism

The notion that alcoholism is a disease with a marked genetic component, and not solely a learned behavior or a coping reaction to stress is a view that has come to be accepted only in the past few decades. A recent Swedish study has identified two distinct types of this disorder. Type I alcoholics, as defined in this study, developed problems with alcohol as adults, experienced a very rapid progression from casual drinking to serious dependency, but were not outwardly abusive or antisocial in their behavior. Type II alcoholics began drinking as teenagers, took much longer to become addicted, but then had serious and recurrent social problems, including conflicts with the law.

The tendency for alcoholism to "run in families" had long been noted, but whether this tendency was caused by something heritable in the genetic sense was for most of this century unclear. Some argued that alcoholism, like "feeble-mindedness" and criminality, was genetically determined and heritable. But others argued that children of alcoholic parents themselves become alcoholic because of behavioral patterns they are exposed to as children and later mimic as adults, or because of the chronic poverty and stress that seem equally locked into certain families.

Insight into possible genetic contributions to alcoholism has come from detailed analyses of family inheritance patterns, and from adoption and twin studies. Children of alcoholic parents adopted into nonalcohol-abusing, unrelated families are much more likely to become alcoholics later in life than adopted children from nonalcoholic parents. The Swedish study just cited examined alcoholism patterns in adopted children. Type I alcoholism had a relatively weak genetic component, probably not more than 30 percent. While some tendency to alcohol use is inherited, type I alcoholism requires some environmental event, usually related to stress, to set it off. Males and females were found to be more or less equally afflicted by this form of the disease. Type II alcoholism, on the other hand, had an estimated genetic component of about 90 percent, with very little contribution from any environmental factors. Type II alcoholics were almost exclusively male. These results may also explain in part why, overall, there are more men than women alcoholics, and why the genetic component for alcoholism appears to be stronger in males than in females. However, within the context of type I alcoholism, there does not appear to be a discernible difference in the modest genetic component between males and females.

There have now been six major studies of alcoholism in twins, and all find the correlation of alcoholism in monozygotic (M2) twins to be much higher than in dizygotic (D2) twins. In a recent twin study carried out at Johns Hopkins University, potential genetic and environmental factors were assessed separately in 113 pairs of male MZ and DZ twin pairs in which at least one twin had a clinical diagnosis of alcoholism. As in most such studies, the number of female twin pairs with at least one alcoholic twin was too small to yield statistically meaningful results. Whereas environmental factors appeared to affect alcoholism similarly in both MZ and DZ male twins, MZ twins showed a much stronger correlation for genetic factors than DZ twins. Environmental effects alone were insufficient to cause alcoholism, whereas genetic factors alone *were* sufficient to cause disease. Genetic factors were highly correlated with age of onset of drinking and a rapid progression to full-blown alcoholism. Environmental factors, where they made a contribution, were correlated with a later onset of drinking and a slower progression toward alcoholism. To date, no twin studies have been carried out using the type I/type II classification of alcoholism, since this is a fairly recent development. A retrospective reanalysis of twin studies already carried out could be done, and this might be very informative.

How alcohol affects neuronal function has always been something of a mystery, since there are no receptors for alcohol on neurons or on any other cell type in the body. Alcohol seems simply to insert itself into the membranes of cells, and in the brain this triggers a sequence of events similar to that initiated by addictive drugs, including stimulation of neurons in the mesolimbic region to release dopamine into the nucleus accumbens. The opioid system is very likely to be involved, since chemicals that block both μ and δ opioid receptors reduce ethanol consumption in addicted rats. Both animals and humans show evidence of increased levels of β-endorphin in the brain after the onset of alcoholism, which could in turn account for the higher levels of dopamine. Drugs that block opioid pathways have also been found to be effective in the treatment of human alcoholism.

One gene that is clearly a factor in the human response to alcohol is the alcohol dehydrogenase gene, which codes for an enzyme of the same name that participates in the breakdown of alcohol in the liver. There is an allele of this gene, quite common in Asian populations, that codes for a somewhat less effective form of the enzyme. The result is an unusual sensitivity to alcohol, including an unpleasant flushing reaction and nausea. The generally negative reaction of many Asians to even small amounts of alcohol is thought to be a major factor in the low incidence of alcoholism among Asian peoples.

The possible involvement of genes encoding neurotransmitters or their receptors in alcoholism has received serious attention in recent years. An initially promising connection between alcoholism and one of the brain receptors for dopamine has generated ambiguous results, but evidence for a connection between serotonin and alcoholism continues to grow. Early onset alcoholism in particular is correlated with low levels of activity in serotonin pathways in the brain, and not surprisingly this correlation is accompanied by more aggressive behavior, including a tendency toward suicide. A particular allele of the gene for tryptophan hydroxylase, a key gene in serotonin synthesis, shows up in a surprisingly large number of these types of alcoholics. The current working hypothesis for these studies is that this allele may result in lower tryptophan production in affected individuals, leading to higher levels of impulsive behaviors and aggression.

Studies in mice confirm an association between decreased serotonin production and an altered sensitivity to alcohol. The 5HT1β receptor knockout mice described in connection with studies of aggression, which have greatly decreased levels of serotonin, were tested for their ability to handle alcohol. These mice drank twice as much alcohol as control mice, actually showing a preference for water containing alcohol compared with control mice. Moreover, they required much higher levels of alcohol to show impairment of behavior. On the other hand, the severity of withdrawal symptoms was not different between the knockout and control mice, indicating that these two aspects of alcoholism may be under separate genetic control. These studies suggest that genetic variations that affect functioning of serotonin (and likely other neurotransmitter) pathways in the brain may have a direct influence on an individual's predisposition to alcohol abuse.*

Numerous other factors could account for the genetically inherited predisposition of some individuals to become alcoholic. Many of these susceptibility factors have also been identified in mice. For example, one obvious difference among individuals that could predispose them to alcoholism is simply a preference for the taste of alcohol. There are inbred strains of mice that differ enormously in their a priori preference for alcohol. When the B6 mice mentioned earlier are given free access to water bottles containing either pure water, or a 10 percent solution of alcohol in water, they will take in 70 to 80 percent of their total water from the alcohol bottle. D2 mice will consume only about 7 percent of their water from the alcohol-laced bottle. This preference is absolutely heritable from generation to generation. Hybrid B6-D2 mice show preference for the alcohol solution intermediate to the two

---

*The involvement of neurotransmitters such as serotonin in multiple aspects of behavior—impulsivity, aggression, substance abuse, for example—deserves some additional comment. The involvement of these neurotransmitters is generally inferred from levels of these substances or their metabolites in spinal fluid or, as discussed here, from gene knockout studies. While such measurements and studies provide strong evidence for involvement of the corresponding neurotransmitter in a given behavior, they tell us nothing about exactly which regions of the brain the behavior is associated with. A global reduction in serotonin function in the brain may well have multiple implications for behavior, but that does not mean that the same region of the brain is responsible for each of the affected behaviors.

parental strains. In Drosophila, normal aversion to alcohol is switched to a preference by mutation in a single chemoreceptor gene.

Another factor in the development of alcoholism, as just discussed, could be sensitivity to alcohol. As the Asian population example shows, it may be insensitivity to alcohol, rather than sensitivity, that predisposes to alcoholism, since those highly sensitive to alcohol are often incapacitated by it at levels too low to trigger addiction. Whatever the explanation may be, it has been clearly shown that sons of alcoholic fathers are at an elevated risk for becoming alcoholics themselves, and as a group they are less sensitive to the effects of alcohol than the sons of nonalcoholic fathers.

When injected with the alcohol equivalent of three to four drinks in humans, most mice will fall asleep. But the length of time they remain asleep before waking up is variable among mouse strains, and is under genetic control. Special strains of mice have now been established that recover quickly from the effects of alcohol (less than 10 minutes), or that take a long time to "sleep it off" (several hours.) The selection curves for establishment of these strains looked very much like those for selection of aggressive behavior (Fig. 9.2), with the gradual separation of sensitivity versus insensitivity taking place over many generations. This suggests that resistance to the effects of alcohol may be a highly multigenic, quantitative trait. Interestingly, mice resistant or sensitive to alcohol tend to have the same sensitivity to other addictive drugs, indicating that genetic predisposition to alcoholism and to drug addiction may share at least some common underlying genes.

The B6 and D2 inbred mouse strains used to study alcohol preference have also been found to differ in their withdrawal response to alcohol. Withdrawal toxicity is a major problem for heavily alcohol-dependent individual humans, and it is a problem fairly specific to alcohol. While withdrawal of any addictive substance can cause extreme discomfort, it is almost never fatal. In advanced stages of alcoholism, sudden withdrawal of alcohol can lead to the state known as delirium tremens, which, if not itself treated, in many cases leads to death. Part of this is due to the invariably poor health of such individuals, but there appear to be specific toxic events involved as well.

The B6 mouse strain, which shows a strong preference for alcohol over water, also shows a much stronger withdrawal response, once it becomes alcohol-dependent, than does the D2 strain, which is less attracted to alcohol in the first place. Both strains can be made alcohol-dependent rather quickly, by injecting them with alcohol and then exposing them to a continuous alcohol-containing vapor for three days. At the end of this time, the mice are monitored for the ease with which a mild physical strain induces them to go into convulsions similar to delerium tremens in humans. The B6 mice showed a much stronger convulsion response than did D2 mice, and at lower levels of alcohol, showing that in these mice alcohol dependence is much more severe.

To get at the identity of the genes contributing to alcohol preference, sensitivity, or resistance, the offspring of B6 mice mated to the D2 strain were scanned for QTL relating to these traits. QTL sites contributing to alcoholism traits in mice have now been identified. Two QTL sites contributing to alcohol preference have been found, one on chromosome 2, and one on chromosome 11. The chromosome 11 QTL is of considerable interest, because it is also the approximate location of the gene coding for the serotonin 1β receptor in mice. Interestingly, these two loci are sex-specific; the chromosome 2 locus controls alcohol preference in females, and the QTL on chromosome 11 affects alcohol preference in males. Seven QTL for sensitivity to the effects of alcohol have been defined, which together can account for greater than 60 percent of the heritability of alcohol sensitivity. A total of eleven QTL controlling the withdrawal response to alcohol have been identified, which together account for nearly 90 percent of the heritability of this trait.

As with any genome scan, the next step is to map the location of all of these QTL more precisely, using more closely spaced markers. When the location of the QTL along the respective chromosomes has been established as closely as possible, researchers will fish out the candidate region of DNA and begin the search for specific genes.

These kinds of studies are important to understanding alcoholism in humans in several ways. First, much more rapid progress can be made doing genome scans in mice compared with humans, and the genes underlying these traits will almost certainly be the same in mice and in humans. The challenge then will be to understand just how it is they contribute to the development of alcoholism. For genes whose identity is somewhat more obscure, we will be aided by the fact that for each mouse chromosome, and even sections of those chromosomes, we know the corresponding human chromosome. So even before the mouse genes are identified we can begin a search for the corresponding genes in humans. A further advantage of the mouse studies is that we can confirm the importance of particular alleles of genes in mice by adding those alleles back into the mouse genome. If an allele of a given gene seems likely to contribute to alcoholism in mice, we can introduce that allele into a mouse not showing that alcoholism-related trait, and watch what happens.

In the meantime, genome scans of human DNA are already in progress. The Collaborative Study of the Genetics of Alcoholism is a six-center program to detect and map QTL, and ultimately genes, whose alleles contribute to alcoholism in humans. Approximately 1000 individuals in 100 or so families have been tested across several generations, correlating the tendency to become alcoholic with inheritance of 291 DNA markers spread throughout the genome. QTL correlating with the development of alcoholism were detected on chromosomes 1 and 7, and a QTL associated with resistance to alcoholism was detected on chromosome 4 (Fig. 11.1). This latter QTL is found in a region of the human genome known to be equivalent to mouse chromosome 3. One of the mouse studies cited above had found a protective QTL at the equivalent position on chromosome 3, but the significance was questionable. However, this locus is very close to the gene for alcohol dehydrogenase, which is important in degrading alcohol in the liver. One of the most intriguing human loci linked to alcoholism occurs on chromosome 6 (not shown here), at a site closely linked to the gene for the 5HT1β receptor.

Progress in both the human and mouse genome scan studies should continue to accelerate through comparisons of information obtained by scanning each genome, and by the increasing availability of information from genome sequencing projects. As with the obesity genome scans discussed earlier, scientists are now at work refining the chromosomal regions containing the alcoholism-related QTL, in preparation for isolation of the corresponding genes. This work will be aided by similar QTL studies in mice, and by completion of the Human Genome Project, which will provide gene sequences for regions of interest.

Current approaches to treating alcoholism, based on a variety of drugs and various forms of psychotherapy, have had a limited impact on this disease. Many treatments can wean addicted individuals from

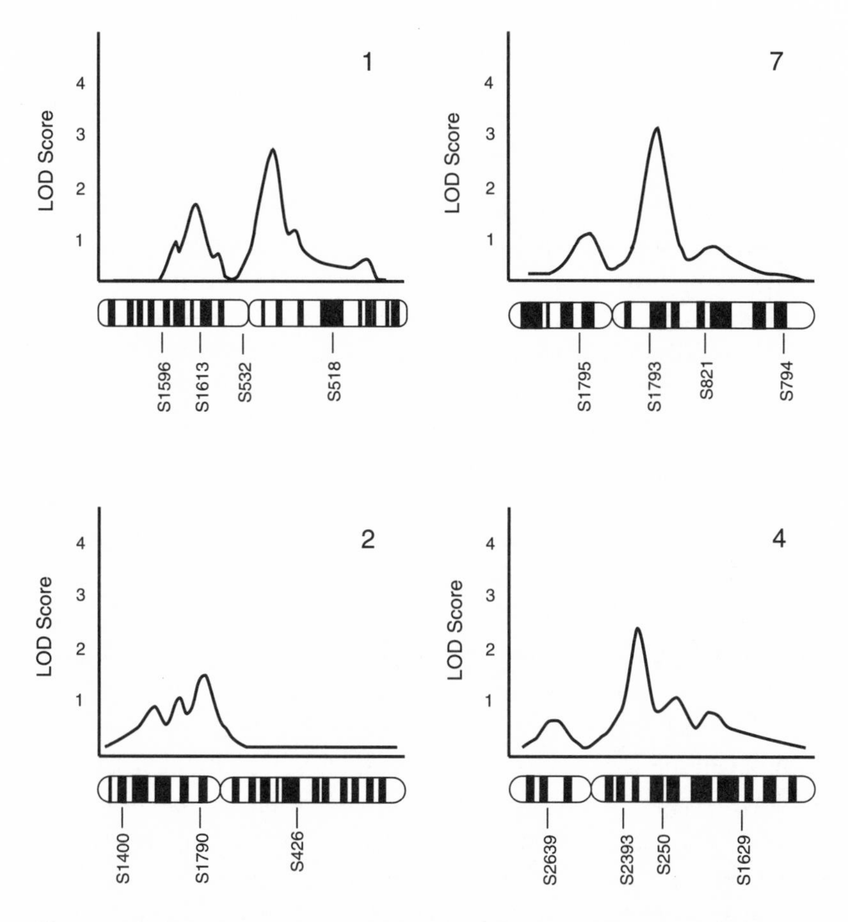

Figure 11.1 Quantitative trait loci relating to alcoholism in humans. Chromosome numbers represented are shown in upper right of each panel. Based on data presented in Reich et al., *American Journal of Medical Genetics* 81:207–215 (1998).

an immediate dependency on alcohol, but the sad fact remains that 50 percent of "cured" individuals revert to alcohol use within three months of the end of treatment. Clearly, new information about the causes of alcoholism is needed if treatment is to be made more effective. Much of this information will come from understanding more fully exactly which cellular and molecular components of the brain and other physiological systems are involved in alcohol dependence. Most important, we need to know how genetically controlled variants in these components affect the alcoholic syndrome, so that treatments can be more intelligently tailored to deal more effectively with predisposing genotypes. Such research is a major goal of molecular medicine.

Genes, Environment, and Substance Abuse

As with all other human behaviors, the substance-abusing phenotype emerges from an interaction of particular genotypes with events taking place in the environment. We are just beginning to perceive what some of the substance abuse predisposing genotypes might include: varying balances among alleles of genetically influenced neurotransmitter systems, endogenous opioids, and possibly hormones, playing against a background of genetic alleles regulating individual sensitivities to the effects of drugs and alcohol. Undoubtedly, as the Human Genome Project progresses, and the results of various genome scans begin to bear fruit, we will discover other genes, perhaps even constellations of genes, predisposing to substance abuse.

But we are already reasonably certain from heritability studies that, after all of the genes involved have been identified, they will be unlikely to account for more than 50 percent of the substance abuser phenotype. The rest of the variability we see in susceptibility to drugs and alcohol will surely stem from environmental factors. Laboratory studies have repeatedly shown that animals reared in environments deprived of social stimulation (cage isolation, with no objects placed in the cage) are much more susceptible to addictive drugs as adults, compared with those raised in enriched environments (cage mates, with novel objects repeatedly introduced into the cage).

Such experiments are obviously not possible with human beings. Moreover, the type of isolation imposed on the laboratory subjects rarely occurs in human families, even in the most deprived environments. Nevertheless, there is ample evidence for environmental influence on addictive behavior. The two most consistently evident environmental factors that seem to push susceptible humans toward substance abuse are substance availability and peer pressure. Someone who, because of genotype, is predisposed to substance abuse, but has no access to drugs or alcohol, obviously will not become a substance abuser. If drugs are available, but one's self-acknowledged peers are uniformly against using them in a self-abusing way, it would be a rare, strong-willed individual who would rise up against such peer pressure and become a solitary, and probably socially isolated, substance abuser.

However, even in the absence of peer pressure and substance availability, we imagine that a skilled parser of personality traits could spot the susceptible individual, because substance abuse is not an isolated personality trait. It occurs in the context of a spectrum of possible personality genotypes predisposing to substance abuse and eventual addiction. We would expect a potential substance abuser to be impulsive, constantly testing the boundaries of the permissible and the possible, perhaps depressed and a bit aggressive. Buried somewhere among these traits, which all evidence suggests reflect particular neurotransmitter balances, we will find the potential substance abuser.

But of course, as we well know, the likelihood of finding an environment completely free of drugs and other addictive substances, with no peers available who would validate substance-abusing behavior, is essentially nil. So we continue to see various susceptibility genotypes tested against the real-world environment, with varying outcomes. Which behavioral patterns lead some individuals to succumb to, and others to resist, developing the substance-abuser phenotype? One factor is certainly peer selection, which is by no means a completely random process. As we have discussed previously, when given a choice, genetically different individuals select from their environment those things with which they are most comfortable, including people with whom they spend time—their peer group. Studies of monozygotic twins reared together or apart show that the twins end up selecting very similar groups of personal friends. Dizygotic twins, on the other hand, select differing groups of friends that by and large reflect their individual temperaments. This manipulation of the environment by individual genotypes makes clean separations of genetic and environmental influences very tricky indeed.

Stress is another environmental factor that seems to be important in expression of a latent substance-abusing phenotype. Some individuals are fairly resistant to stress, and in general are more "centered"—resistant to environmental influences. Others are more vulnerable to pressures from the surrounding world. This, too, is a reflection of the ongoing dance between genotype and the environment in producing a particular phenotype. We may never be able to separate the dancers completely; perhaps that should not even be our goal. The important thing is for those treating substance abuse, and for those whose lives are intertwined with substance abusers, to understand that this is a problem played out against a background of genetically influenced personality and behavioral components. The more we learn about how these components work at the cellular and molecular level, the more likely we are to be able to influence them through medical interventions.

It is equally important for those enamored of a cellular and molecular approach to behavior to understand that no matter how completely we identify all of the genes influencing a given behavior, the data tell us we will rarely if ever account for much more than half of the variability we see in that behavior among human beings. The rest is the domain of the behavioral psychologist. And no matter how precisely we define the function of any behavioral gene at the physiological level, this function gains meaning only when the gene is set in motion within the context of a particular environment. That is why behavioral genetics is so complex. But it is only by acknowledging the distinction between genes and environment, and by understanding what both behavioral genetics and behavioral psychology have to offer, that we will find the synergy between these two approaches that may someday allow us to deal effectively with something as complex as substance abuse.

12

The Genetics of Human Mental Function

There is little question that the lives of cells are governed closely by genes. The qualities of the proteins encoded by various alleles of individual genes present in single-cell organisms determines every aspect of their ability to survive and reproduce. Among single-cell organisms such as yeast and Paramecia, the effects of single genes on the activities of these cells as they go about the business of reproducing, that is, on their behavior, are relatively easy to detect. In multicellular animals, where the many functions constituting behavior are distributed among the somatic cells making up various tissues and organs, the role of individual genes can also be discerned.

It is precisely when we come to the cellularly more complex nervous systems of larger multicellular animals that the issue of genes and behavior becomes complicated. The major complication arises because, more than in any other organ or system in the body, the behavior of cells in the nervous system can also be affected by the environment. Neurons are our window onto the world around us. We use the various images provided by sensory neurons to formulate responses to our environment, and this experience of our environment, and our responses to it, are remembered. Neurons are altered by contact with the environment, in ways that are still only crudely understood, but which involve both intracellular chemical changes and physical alterations in the synaptic connections they make with other nerve cells. These "conditioned neurons" alter the way we respond to the same information when it appears in the environment again.

Beyond doubt, the mechanisms used by neurons to carry out their functions are also under the direct control of genes. Having one or another allele of a key gene regulating neuronal function could very well have a major impact on how a neuron perceives environmental events, how rapidly it responds, or how aggressively it seeks out new connections with other neurons. We have now seen numerous examples of the ways in which variations in neurotransmitters and their receptors can affect behavior. But the fact that the environment, as well as genes, can affect the way neurons themselves behave introduces an important element into the behavior of multicellular organisms. Just as the behavior of individual neurons is altered by experience, so too is the behavior of the whole organism. Determining the relative impact of genes and environment on the behavior of neurons and individuals is at the very heart of the "nature-versus-nurture" controversy in the analysis of human behavior. And nowhere is this issue more controversial than in the analysis of human mental function.

That human mental function under a given set of conditions is variable among individuals must have been noted since the beginning of organized human society. For virtually every other trait in our bodies, there exists a great deal of variation in the population as a whole in terms of structure or function. People are tall or short, dark-skinned or light, timid or assertive. From a biological point of view, we imagine that the purpose of all this variation is to provide a pool of resources that we as a species can draw upon should environmental conditions change suddenly. The precise phenotype that might survive best during the coldest part of an ice age may be different from the phenotype optimal for a warmer period. Maintaining a range of phenotypes (and of the underlying genotypes) could give a species a better chance of surviving sudden environmental challenges; the advantage of this to the individual is that it assures sexually reproducing members of that species of an adequate pool of breeding partners for dissemination of their own DNA.

The question we must ask ourselves is whether variability in human mental function is any different in this respect. Certainly, it is one of our most valuable tools in dealing with the world around us. It seems entirely reasonable that in our evolutionary past, different ways of thinking, different mental responses to the environment, may have had different survival values at different times. It may also be the case that at any given instant in time, having a range of mental traits within the species—linear versus creative, for example—helps the species as a whole to better meet environmental challenges and maintain a vigorous breeding pool. But to what extent are these different ways of thinking affected by our genetic inheritance? One could imagine two possibilities defining the extremes of this question. It could be that variabilities in every aspect of mental function, from perception and interpretation of the environment through storage and retrieval of memory, from whittling a stick to writing a symphony, is dictated entirely by variations in our genes. Alternatively, one could imagine that genes control the structure and operation of neurons, but that variability in mental function is determined entirely by the differing sets of experiences imprinted on these neurons (and the individual) by the environment.

Before we examine the evidence relating to this range of possibilities for defining human mental function, we should say something about a word we have avoided using so far: intelligence. It would be nice if we could avoid doing so altogether; it is perhaps one of the most sensitive words in science and society today. Uncertainty about just what is meant by intelligence has generated a great deal of heated debate, and has slowed research into the genetic basis of human mental function for the better part of a century. No other question so intimately affects the essence of what it means to be human. We must always remember that each of us brings to this debate a history of personal, cultural, social, and political preconceptions about the underlying issues and their broader implications.

In this book we take the point of view that, from a scientific standpoint, and probably from a societal point of view as well, there simply is no adequate definition of human intelligence. Current definitions of intelligence are most often tied into learning ability. But the ability to learn what? Music? Math? Painting? Survival in the woods? Intelligence is also often tied to verbal and reading skills. But reading what? Words? Music notes? Math symbols? Footprints in the snow? Each of these represents a quite different type of mental function, and they are combined to different degrees in different people. Standardized tests to measure so-called intelligence measure largely mathematical, reasoning, and verbal skills. What do these tell us about creativity, the ability to start a business, or the insights necessary to optimize human relationships? The problem is, from a scientific point of view, we know absolutely nothing about the biological significance—the survival value in the wild under differing conditions—of any other of these forms of mental functioning. We interpret them within the context of our own contemporary society, but that is not where they evolved. We can try to make evolutionary sense of the variability we observe in what is now called human intelligence. In a defined set of environmental circumstances, we could possibly even point to one or another combination of mental traits that might be advantageous. But beyond that science simply cannot go.

So in this book we do not use the term intelligence, except when necessary for historical context; it is far too laden with emotional meaning, and as such means different things to different people. That there is variability in the way people think about and solve problems is evident from any number of standardized tests. Whether or how this can be related to something we could call intelligence does not concern us here. What we are really interested in is whether and to what extent this variability is heritable, and therefore likely to have a genetic basis. To avoid obscuring this goal, we refer to the way human beings think and solve problems simply as human mental function.

Is There Heritable Variability in Human Mental Function?

One approach to resolving the question of the relative role of genes versus environment in determining human mental function is to study its heritability. The logic behind this approach is simple. To the extent that the apparent variability in human mental function is determined by allelic variants of genes underlying the mental processes involved, then transmission of these alleles through the germline should be discernible in family studies. This implies, of course, that we know how to measure mental function, how to sort it out from all the other traits that go into making up a human being. We return to this important question below. Let us also note that, given the complexity of human neurological function, we would not expect to see many instances of single genes explaining an entire mental function. Alterations in genes involved in mental processes that render a gene product unable to function may cause grossly observable pathologies, and in some cases the underlying gene can be isolated and identified. But we have already learned to view such findings cautiously; just because loss of a gene cripples a particular function does not mean the gene explains the function.

We fully expect that human mental function will involve large numbers of genes working together in as yet undefined ways. Still, given the power of modern molecular genetics, we could hope eventually to be able to follow genes—even groups of genes—relevant to mental function through hereditary pathways. On the other hand, that portion of mental function that is a product of environmental imprinting on neurological pathways should *not* be heritable. The unique set of environmental experiences accumulated by an individual will be extinguished along with that individual at the end of life, just as increased adventurousness induced in nursing mice or rabbits by increased environmental stimulation is not passed on to subsequent generations. Neurologically acquired experience cannot be transmitted through the germline. Acquired experience can of course be passed from one generation to the next through cultural transmission, but at least in theory that can be distinguished from genetic transmission by studying cases in which newborns are transferred from their birth culture to a different culture before culture can be imprinted.

The first serious proposal that the variability we see in mental traits might be inherited in the same way as any other human biological variable is generally credited to Francis Galton. Galton was a cousin of Charles Darwin, and he was greatly impressed by his cousin's thinking about the meaning of variability in terms of the evolution and heritability of diverse biological characteristics. Both Galton and Darwin remarked on the clear heritability patterns of behavior in animals such as dogs. Galton, however, was the first to suggest that variability in human behavior, and particularly in the way individuals think in different circumstances, ought not to be different from any other trait in terms of heritability, a point of view that had escaped Darwin initially, but which the latter quickly embraced. Galton also felt that, to the extent these differences are heritable, it ought to be possible to alter the overall phenotypic composition of the human species by selective breeding practices.

Galton did more than just ruminate about the heritability of mental traits. He took the first halting steps toward developing the science now known as behavioral genetics.* He began the systematic collection of information he felt would be required to substantiate or refute his ideas. Doubtless he was impressed by the enormous amount of information Darwin had collected during his *Beagle* voyage and elsewhere in support of his ideas on basic evolution. Galton's own assemblage of information is somewhat bewildering in its range and complexity. His initial standards for assessing an individual's mental capacity were rather crude and off the mark; he assumed, for example, that a person's social position in British society was an accurate reflection of that individual's intellectual prowess. His early attempts to quantitate mental ability focused largely on neurological parameters such as speed in recognizing certain objects, ability to memorize particular words or numbers, reflexes, physical strength, and so on.

Galton shared with Gregor Mendel, the Austrian monk whose work with plant breeding set the foundation for the field of genetics at the dawn of the twentieth century, a passion for quantitation. This was not generally the case for nineteenth-century biologists, and it took many years for the contributions of both to be fully appreciated. Galton recognized that Gaussian distribution functions, originally derived to deal with errors in measurement, could also be used to analyze variabilities around a mean for virtually any parameter, including the distribution of human phenotypes. His work set the stage for the development of statistical treatments of a wide range of human traits, and eventually provided the tools of choice for analyzing heritability in long-lived populations such as humans, where controlled breeding experiments are (for many reasons) out of the question. Galton also

* Genetics as a formal science was not yet developed in Darwin's and Galton's time; the notion of genes as the units of inheritance was established only in the first years of the twentieth century.

shared with Mendel the fact that neither was caught up in the intense discussions of many of the exciting scientific issues of their day, and neither was part of the larger community of individuals moving biology in new directions. This distance between each of them and their respective scientific peers contributed significantly to an initial underappreciation of their work.

Galton was the first to introduce the phrase "nature versus nurture," and although he clearly favored the former as an explanation of human behavior, he had little direct evidence to support his bias. Nevertheless, Galton's contributions to human genetics, most notably his insistence on statistical interpretations of large amounts of population-based data, were seminal to the field. The methods he developed, admittedly with substantial subsequent modification, are those still used today for studying heredity, and hence the role of genes, in human populations. He was also among the first to recognize that the offspring of parents representing one extreme or the other of a given trait within a population tend to be more like the average for the population than do their parents. This "reversion to the mean" led Galton to doubt for awhile the possibilities of selective breeding as a means of altering human phenotypes. Galton was also the first to recognize the unique possibilities offered by the study of twins in matters relating to human genetics, although he never really pursued this possibility.

By the end of his studies Galton was convinced that most human traits are indeed heritable. Although contemporary scientists would be unlikely to find his results compelling in terms of the data he had at hand, his various publications on the subject stimulated others to take up where he had left off, and the field staggered forward. Galton set forth his thinking in two major works: *Hereditary Genius: Its Laws and Consequences* (1869) and *Inquiries into Human Faculty and Its Development* (1883). It was in the latter work that he introduced the term eugenics, which he defined broadly as social intervention to promote breeding practices that would favor the generation of superior individuals.

And therein lay the seeds of one of the most disturbing controversies ever to arise out of biological research. Galton was not the first to propose selective breeding of particularly "fit" individuals as a means of improving society; Plato proposed much the same thing in his *Republic* over 2000 years earlier. The difference was that by Galton's time, we were beginning to understand enough about heredity to make such a proposal seem at least theoretically feasible. What happened in the field of eugenics in the first part of this century has affected the field of human genetics profoundly, and has made society at large and even many scientists suspicious of those who would tinker with human genes. No discussion of human genetics would be complete without looking back at what happened. A brief history of the eugenics movement is given in Appendix II.

Among most eugenicists, attention became focused to an excessive degree on heritability of mental function, particularly as revealed by so-called IQ tests. While there was a great deal of discussion among lay supporters of eugenics about the heritability of traits such as criminality or alcoholism, scientists focused their attention on performance on IQ tests, mostly because they had what they felt was an objective tool for measuring it. The IQ tests available in the first third of this century were certainly crude, and not particularly meaningful. Politically inspired claims that particular ethnic groups as a whole performed poorly on these tests were initially seized upon to justify a wide range of xenophobic undertakings. But such claims were eventually put to rest by studies showing that within a generation or two test scores among the descendants of these ethnic immigrants were indistinguishable from "mainstream" scores. Even the most diehard eugenicists were forced to admit that these changes could not be explained on a genetic basis. Other studies showed the impact of inadequate nutrition and poor language skills on test performance. The clear cultural bias of most IQ tests, referring to things or to concepts with which the test taker may have had no familiarity, was argued well into the 1970s. But other studies continued to show that, although the level of performance on IQ tests was not significantly different among racial or ethnic groups once they had entered the mainstream of American society, the variability of test scores *within* these groups was as wide as the variability among mainstream groups. This should not be surprising. The number of traits we use to define race—based entirely on observable physical characteristics—is very small, and the likelihood that many other traits would be linked to this limited set of characteristics is vanishingly small.

As originally conceived, IQ tests were designed to detect areas of weakness in what students were learning in school, so that appropriate corrective measures could be made to address specific deficits. Only later did these tests come to be used as a measure of general intelligence, rather than accumulation of knowledge. Initially, IQ was defined as the ratio between the age level of performance achieved on a given test and the test taker's actual age, multiplied by 100. For example, a six-year-old performing at a nine-year-old level would have an IQ of 150. Since this would not be a useful calculation for adults, IQ scores today are based on how an individual performs in relation to his or her age group in the population a whole. Moreover, in order to deemphasize accumulated knowledge in favor of generalized mental ability, more sophisticated test instruments have been devised that probe a wide range of human mental functions. The components of one of the most commonly used tests, the Wechsler Adult Intelligence Scale (Revised), are shown in Table 12.1.

The debate about whether IQ tests measure acquired knowledge (content) or basic mental function (ability) continues to this day. At the same time, evidence that variability in performance on IQ tests is

Table 12.1. *The Wechsler Adult Intelligence Scale (Revised)*

Areas tested	Presumed underlying ability
Verbal	
Information	Stable knowledge
Digit span	Short-term memory
Vocabulary	Reading and word skills; long-term memory
Math	Mathematical reasoning
Comprehension	Social awareness
Similarities	Abstract reasoning
Performance	
Picture completion	Visual reasoning
Picture arrangement	Visual skills
Block design	Visual-motor abstract ability
Object asembly	Integrative reasoning
Digit symbol	Visual-motor coordination

heritable—whatever it is these tests actually measure—continues to grow. As with the relation of genes to human personality, the best data we have for estimating the relative contributions of genes and the environment to mental function comes from family studies, particularly studies of adopted children, and of monozygotic and dizygotic twins reared together or apart. A few starts in this direction had been made in the 1930s; the results were suggestive, but not definitive, and were understandably greeted with suspicion by those becoming alarmed about the social and political implications of eugenics. But a comprehensive reanalysis carried out in 1960 of the existing literature on adoption and twin studies revealed a definite correlation between degree of genetic relatedness and performance on IQ tests. For example, a parent and an adopted child are no more likely to score the same on an IQ test than two randomly selected unrelated individuals, even after many years of a close family relationship. A parent and a biological child, on the other hand, are much more likely to score close to one another—even when the child is adopted out at birth, and the parent and child are tested only later in life.

Follow-up studies over the ensuing years have strengthened this correlation. Detailed analyses of monozygotic twins reared together or apart, such as those from the Minnesota Twin Study, have been particularly informative. Thomas Bouchard, in a paper published in 1998, summarized the results from five different, well-controlled twin studies (Fig. 12.1). Genetically identical twins reared together or apart had IQ test score correlations of about 0.75; dizygotic twins reared together or apart had IQ test score correlations of 0.38. Adopted children and their adoptive siblings had IQ test correlations of about 0.28 when they were young, but this correlation essentially disappears by the time the children leave home. According to Bouchard, these correlations allow the conclusion that at least 70 percent of the variability among individuals in performance on IQ tests is due to genetic differences. The results in all of the twin studies, using a variety of testing techniques, were remarkably similar. So while we cannot be entirely sure exactly what aspect of mental function it is that IQ tests measure—perhaps nothing more than the ability to take IQ tests—it seems quite clear that variability in performance on these tests is to a large extent heritable, and thus has a genetic basis.

Two rather startling conclusions have emerged from the twin analy-

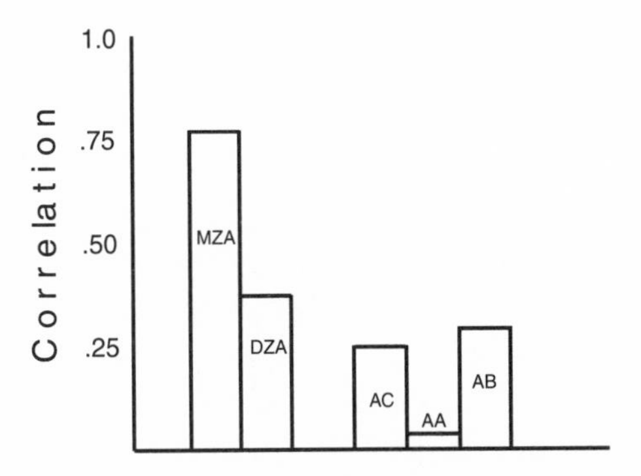

Figure 12.1 IQ test correlations for monozygotic twins reared apart (MZA); dizygotic twins reared apart (DZA); between adopted children and their adopting siblings, as children (AC); between adopted children and their adopting siblings, as adults (AA); and between adopted children as adults and their biological parents (AB). Based on data presented in T. Bouchard, *Human Biology* 70:257–259 (1998).

ses, and from the study of biological versus adopted siblings. First, there is an increasing influence of genetic background on IQ test performance with age; second, shared environmental influences during childhood account for very little of the IQ test performance variability seen in related children reared together or apart.

Both of these findings seem counterintuitive. We would expect that the experience and wisdom gained over a lifetime must contribute markedly to general mental function, and come to play an increasingly important role in performance on IQ tests as life goes on. The notion that enriching the environmental input individuals receive throughout life will improve their ability to function in society is the basis for any number of educational and social programs. These programs may in fact achieve those aims, but the data suggest that this is not through improvement of any of the parameters measured by IQ tests. Several studies have now shown that in fact, as individuals get older, genetics actually increases as an explanation for IQ test performance, just as it does for personality. In very old individuals, the influence of genes may fall off somewhat, but this is doubtless confounded by the increasing (and generally unmeasured, because it is unmeasurable) incidence of preclinical senile dementias.

In IQ testing, as in personality assessment, psychometricians generally recognize two sorts of environmental influence when examining heritability of test performance within the same family, regardless of degree of genetic relatedness. Shared environmental influences are those things to which all children are exposed equally: the parents themselves; going to the same church or school. Nonshared environmental factors are things each child does on his or her own; being or not being in the scouts, for example; taking or not taking music lessons; selecting peer group members. As with the development of human personality, manipulation of or selection from the shared environment may be dictated to some extent by each individual's genetic makeup. In the case of genetically identical twins, there is also concern among some researchers about the possible long-term influence of different placental environments, both within the same placenta, and for twins having separate placentas within the same womb.

Comparisons of how identical twins reared apart, and unrelated children reared in the same home, perform on IQ tests show at best a modest effect of shared environmental factors on test performance. Studies that follow these children over time, such as the Colorado Adoption Study, suggest a modest effect of a shared home and school environment when children are quite young, but by the teen years this effect largely disappears. This conclusion is also supported by the data shown in Figure 12.1. Children adopted at a young age may mirror to some extent the IQ test performance levels of their adoptive parents and siblings for a few years, but by the time they leave home their IQ scores are closer to those of their biological parents, whom they never knew, than to their adoptive family. To the extent that environment is a factor in IQ performance, the preponderance of the data shows that children's peers are a greater influence than are their parents or siblings.*

The bottom line is that, with the most sophisticated analyses presently available, virtually none of the variability in the IQ test performance of siblings as adults can be traced to shared environmental experiences, although a modest portion can be traced to nonshared influences. Nor can the IQ test performance of biological siblings raised in different homes be correlated with the educational level or socioeconomic status of the rearing family. The data on IQ testing mirror the

---

* For a further discussion of why this may be so, see the book by Judith Rich Harris, *The Nurture Assumption: Why Children Turn Out the Way They Do* (New York: Free Press, 1998).

results discussed earlier on the development of human personality; if anything, the genetic influence on IQ test performance is stronger. As with personality, these findings seem counterintuitive. Do none of the efforts of parents to educate their children matter? Surely they do. Education is largely a matter of accumulation of cultural information. Exposure of children to rich sources of cultural stimulation will facilitate that process. But IQ tests are not designed to measure accumulation of cultural information. In the early years of IQ testing, they did do that to a large degree, and they were rightly criticized for that fact. To some extent they do so to this day; the tests are administered in a given language, and language is cultural. So are dialects within a given language. Children speaking different dialects, or with innate or acquired language difficulties, will perform differently on IQ tests from children without such differences or difficulties. But the more these tests are refined to eliminate or compensate for these inherent ambiguities, the more they emphasize the fact that shared environmental influences have a minimal impact on the variability in test performance.

Beyond doubt, we can help our children optimize their mental abilities through cultural enrichment, and by providing a warm and protective environment in which they can explore and use their own native mental abilities. But we impart those abilities to them through our genes, and we cannot alter our children's genetic endowment. We can greatly impair their mental development, however. Extreme cases of cultural deprivation during critical periods of neurological development, as seen in the child called "Genie" who was kept locked in a closet for twelve years,* can lead to severe emotional damage and even mental retardation. We can destroy our children's natural mental abilities, but we cannot improve them. We may wish it were different, and intuitively we may think it should be different, but at the present time there are simply no data suggesting that it is. We can provide a secure and culturally enriched environment that will allow each child to optimize his or her innate abilities, but we cannot fundamentally alter these abilities.

---

* R. Rymer, *Genie: A Scientific Tragedy* (New York: Harper Perennial, 1993). This is also portrayed dramatically in the film *The Apple*, by the young Iranian director Samira Makhmalbaf.

Genes and Variability in Mental Function

So how and where do genes come into all of this? What are the genes that account for the heritable variability in human mental function? As with all human behavioral genetics, the means we have to answer such questions is limited. One approach is to look at naturally occurring alleles of genes within the population that are known to cause deficits in mental function. Once the gene and its product have been identified, we can infer their involvement in normal mental processes. Alzheimer's disease is often cited as an example of a major mental disorder brought about by naturally occurring alterations in individual genes, but it is in fact an example of how such mutations may have absolutely nothing to tell us about the normal processes involved in the underlying behavior. For that reason, it is worth taking just a moment to look at the genetics of this devastating disease.

Based on the timing of onset of the disease, and to some extent on the underlying defect, Alzheimer's falls into two distinct categories. Approximately 95 percent of cases are *late-onset* Alzheimer's disease, which sets in after sixty years of age. About 5 percent of Alzheimer's cases are of the inherited type, and are characterized by early onset of disease symptoms—often well before sixty years of age. Yet even this early form of Alzheimer's occurs well after the reproductive period of an individual's life, allowing the altered genes involved to be passed on to the next generation.

Early-onset Alzheimer's disease is caused by mutations in one of several possible autosomal-dominant genes; inheritance of a single altered allele is sufficient to confer susceptibility to disease. One of these genes has been traced to chromosome 21 and is called *app*, because it encodes a protein called *amyloid precursor protein*. This protein is the precursor form of the amyloid protein found deposited in neuritic plaques—regions of destroyed brain tissue characteristic of Alzheimer's disease.

But mutations in the *app* gene turn out to account for only a minor portion of early-onset disease. By far the majority of cases involve mutations in two additional genes, discovered in recent years, called presenilin-1 (*ps-1*) and presenilin-2 (*ps-2*). Over thirty mutations have been found in *ps-1*; only two have been found so far in *ps-2*. The *ps-1* mutations account for roughly three-quarters of early-onset cases. Disease induced by *ps-2* is more rare, and appears to be milder and slightly later in setting in. We do not yet know the precise functions of the *ps-1* and *-2* gene products, but the net result of these mutations is the same as the *app* mutation: production of abnormal amyloid protein.

The genetic basis for the late-onset forms of Alzheimer's disease is less clear. So far there is little evidence for an involvement of the three genes responsible for early-onset disease. For several years now, it has been suspected that a gene located on chromosome 19, called *apoE*, may be involved. The ApoE protein is also part of the complex called LDL (low-density lipoprotein), which helps transport cholesterol through the blood. There are three different alleles of the *apoE* gene in the population, called ε2, ε3, and ε4 (see Table 12.2). Some 60 percent of the population in the United States is homozygous for ε3 (i.e., is ε3/ε3). Among late-onset Alzheimer's patients the frequency of this allele is twice as high as in age-matched control groups. Statisticians have calculated that someone with two ε4 alleles (i.e., someone who is ε4/ε4; 2 to 3 percent of the U.S. population) is three times as likely to develop late-onset Alzheimer's disease as someone with no ε4 alleles.

None of the genes identified as causing Alzheimer's disease is likely to play a direct role in normal mental functioning. In early-onset disease, all of the gene defects result in the buildup of amyloid deposits, which could conceivably inhibit the function of surrounding neurons. It is obvious that anything leading to the death or dysfunction of entire neurons will lead to grave mental disturbances, but the genes causing death or dysfunction are highly unlikely to have anything to do with the affected mental processes. The same is true for the *apoE* gene of late-onset Alzheimer's disease. The most likely function for ApoE is in protecting against oxidative damage and deposit of harmful plaques in blood vessels. But the likelihood that the *apoE* gene plays any direct

Table 12.2. *ApoE allele frequencies in patient and control populations*

	ε2	ε3	ε4
Control population	.08	.77	.22
Early-onset Alzheimer's	.06	.61	.33
Late-onset Alzheimer's	.03	.51	.46

role in the mental functions gone awry in late-stage Alzheimer's disease is just about nil. So we must be very careful in extrapolating from the identification of a gene associated with a disordered behavior to a role for that gene in the behavior per se.

Several other genes have been identified, through the study of specific cases of mental retardation, that appear more promising than the Alzheimer's genes identified so far. One of these, *fmr2*, is associated with a form of retardation called the "fragile X syndrome." Like many other such genes identified, *fmr-2* is on the X chromosome (the fact that the X chromosome is not paired makes inheritance of genes on this chromosome easier to detect). The product of *fmr-2* regulates the expression of other genes in cells, making this gene an excellent candidate for normal neuronal function. Other X-associated genes have been identified as causing mental dysfunction, and some of these are involved in a direct way in synaptic function or, like PDE-4 and CaMKII, in intracellular signaling pathways known to be important in neurons. Almost certainly, unraveling the precise functions of these genes will provide important insights into how mental processes are carried out in the brain.

Somewhat surprisingly, there have been very few specific correlations at all of mental function, judged by any criterion, with neurotransmitter function. Global deficits in neurotransmitter pathways in the brain are often correlated with diminished mental function, but it is rarely clear in such cases whether the neurotransmitter deficit is a cause of the mental deficit, or simply a concomitant of it. Generally, the neurotransmitter problem is so severe that it would be surprising if mental function remained unaffected. As discussed earlier, loss of the *maoa* gene, which controls the levels of several neurotransmitters through the synaptic reuptake mechanism, leads to mental retardation, at least in the one family in which this has been studied. But in general, there are no quantitative correlations of mental function with particular *functional* alleles of genes governing neurotransmitter systems.

One mental function commonly measured by instruments such as the Wechsler Adult Scale is the ability to make connections between disparate pieces of information under the pressure of time. From that point of view, it will be exceedingly interesting to follow investigations in humans of the gene affecting the NMDA receptor in mice, as described in Chapter 8. The ability to make associations quickly, accurately, and effectively is certainly a useful skill, and one that may underline many interpretations of intelligence. The mouse studies also showed that once the genetic change was made in the germ lines of these mice, the measurable trait was faithfully inherited from generation to generation. This certainly reinforces the notion that genetic variations affecting mental function can be heritable.

Despite some successes, looking for individual genes associated with mental function may be an inherently inefficient way to go after the genes involved in normal mental functioning. Most genes that have been found in this way, like those involved in Alzheimer's disease, have in fact turned out to be genes that cause gross alteration or destruction of entire cells, or even blocks of cells, and such genes provide no insight at all into how brain cells function internally or with each other.

An alternative approach has been taken in a joint U.S.-British project headed by Dr. Robert Plomin. In this project, called an allelic association study, the DNA of individuals who perform differently on IQ tests is tested for particular alleles of polymorphic genes or DNA markers* already known to be associated with, or closely linked in DNA to, neurological function. DNA samples have been collected from hundreds of individuals, and pooled into several groups that score high, medium, or low on standardized IQ tests. These DNA pools are then analyzed for the presence or absence of particular alleles of some 90 known candidate genes or DNA markers. This approach differs from a standard genomic scan in that in a genomic scan, one carries out inheritance studies to look for candidate genes associated with a given trait with no foreknowledge of what those genes might be. In Plomin's approach, one begins with genes or markers already identified as possible candidates for involvement in some level of neurological function, and asks whether this inheritance correlates with inheritance of a particular allele of any of the candidate genes or markers.

To date, no definitive gene associations have been uncovered in these studies, but it is highly likely that some will be. The same approach, whether in the form of genome scans or association studies, is already yielding results in fields as diverse as alcoholism studies and

---

* The nature and use of DNA markers is discussed in more detail in Appendix I.

obesity. The yield from such studies will very likely accelerate as the Human Genome Project approaches completion in the first few years of the next century. From what we know so far, we might reasonably expect that the genes governing performance on IQ tests will be no different from the genes discovered on the basis of their involvement in other nervous system functions—largely genes that affect the rate and direction of exchange of neuronal signals. The likelihood that we will discover anything even remotely identifiable as "intelligence genes" is just about nil.

13

The Genetics of Human Sexual Preference

From time to time a report appears in the scientific literature that forces us to pay renewed attention to a trend in thinking that had been developing all along, but which, until the report was published, we could avoid thinking about. Such a paper appeared in the journal *Science* in July 1993. Summarizing several years of work by the laboratory of Dr. Dean Hamer, at the National Cancer Institute, the report was entitled "A Linkage between DNA Markers on the X Chromosome and Male Sexual Orientation." In it, Hamer and his colleagues presented the first serious evidence at the molecular level that male homosexuality has heritable components, and that at least one of the genes underlying this trait is transmitted exclusively through the maternal parent. The response, both from the public and the scientific community, was almost instantaneous. The report was variously praised, rejected, feared, welcomed, or damned, depending on the audience. But it was not ignored.

The notion that sexual orientation might have a significant genetic component did not originate with the Hamer paper. The tendency for homosexuality to run in families had been noted for over sixty years. Many of the early studies were small and somewhat anecdotal, but more recent, larger studies have confirmed this general finding. A study published in 1986, for example, found that 22 percent of the male siblings of self-described gay men also described themselves as gay. That is well beyond what we would expect on a chance basis, since not more than 5 percent of the male population in the United States is estimated to be gay. Indeed, in the same study only 4 percent of the male siblings of an equivalent group of heterosexual males described themselves as gay. Overall, the frequency of gay brothers is three to six times higher for gay than for heterosexual males. It had even been noted in a few of these early studies that the trend toward male homosexuality appeared to run through maternal rather than paternal lineages. Homosexual behavior in females also tends to cluster in families, although there have been fewer studies and the data are less strong.

It had also been recognized that male homosexuals are more likely to have a gay brother rather than a lesbian sister, whereas lesbians are more likely to have a lesbian sister than a gay brother. In other words, male and female homosexuality tends to run in separate families. That was one piece of evidence suggesting that, to the extent homosexuality would be affected by genes, the genetic programs might not be entirely the same in men and women. Gay men and women also differ behaviorally. Males tend to be exclusively heterosexual or homosexual, and are only rarely bisexual. Women show a much broader spectrum of preferences, with a much higher percentage of nonheterosexual women showing varying degrees of bisexuality.

For those who did not want to believe in a genetic basis for homosexuality, the family studies could be interpreted on the basis of a common environment that might predispose to homosexual behavior, particularly in the sense that one homosexual son or daughter could tempt other siblings into the same sort of behavior. The twin studies were harder to dismiss. In seven studies of monozygotic and dizygotic twins, it was found that in 244 monozygotic twin pairs where one twin was homosexual, the other was homosexual as well 58 percent of the time, on average (Fig. 13.1). In 175 dizygotic twin pairs, both were gay 18 percent of the time, which is very close to the figure for nontwin siblings. Where it has been looked at in large enough numbers of subjects, these concordances were true of female twins as well as males, and of twins reared together or apart. One study included three sets of triplets in addition to monozygotic and dizygotic twin pairs. One triplet set contained a pair of monozygotic male twins, both of whom were gay, and a sister who was not. A second mixed set had two monozygotic female twins, both of whom were gay, and a heterosexual sister. The third set were all monozygotic males, and all three were gay. In a study of 57 male infants adopted into families in which at least one male child turned out to be gay, the adopted male child was also gay in 11 percent of cases.

The high incidence of both monozygotic twins being gay, particularly in those cases where they were reared apart, raised a serious challenge to either the family environment or the environment in general being the sole factor in determining a homosexual lifestyle. Analysis of the available twin data suggest the genetic component of homosexual behavior is about 50 percent. On the other hand, the finding of 11 percent homosexuality developing in adopted siblings, which is considerably higher than the 5 percent or so estimated for the population at large, raises some questions. This effect could be only apparent, due to the relatively small number of individuals involved, but it could also reflect an influence, direct or indirect, of siblings on one another in the determination of becoming gay. We could imagine, for example, that

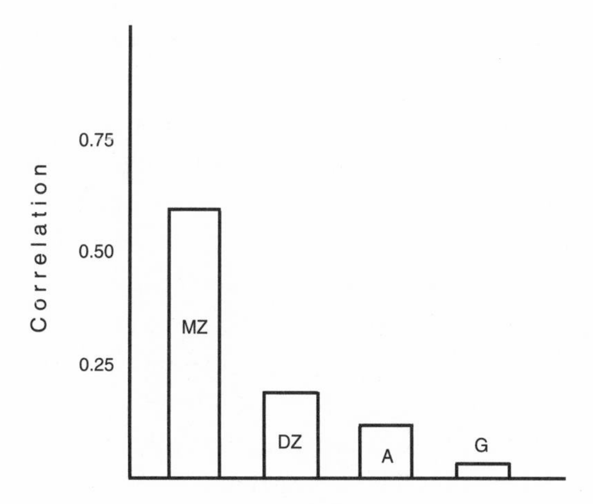

Figure 13.1 Sexual preference correlations in twin pairs (MZ, DZ), between adopted children and adoptive siblings (A), and in the general population (G).

an individual with only a modest genetic predisposition for becoming gay could find that propensity strengthened in a home where a close sibling followed a clearly gay developmental pathway.

While the family and twin data suggest that homosexuality is a heritable trait, none of the data suggest the existence of anything we could call a definitive " gay gene" which, if inherited, would cause an individual to be homosexual. Indeed, the family and twin studies make it clear that homosexuality, like any other behavioral trait, is not Mendelian (single-gene) but quantitative, almost certainly involving multiple genes. Like any other behavioral trait, there is also a substantial contribution of nongenetic factors, such as environment. On the other hand, the fact that environment rarely accounts for more than 50 percent of homosexuality makes it clear that environment is unlikely to be the sole determining factor. Nevertheless, even with the twin and adoption studies, those opposed on political, social, or moral grounds to the idea that homosexuality has a substantial genetic component could still find room to quibble.

Hamer's study, if confirmed, makes such quibbling more difficult. Hamer and his research group focused on two questions. First, they wanted to determine whether the apparent tendency for male homosexuality to pass through maternal lineages was true. Males have only one X chromosome (they are XY with respect to the sex chromosomes), and their single X chromosome is always and only inherited from the mother. In males, altered allelic forms of genes on the X chromosome may cause the appearance of a trait that is not seen in females inheriting exactly the same chromosome. Hemophilia and certain immune deficiency diseases are examples of X-linked traits that show up almost exclusively in males. The reason is that females have two X chromosomes (e.g., they are XX), and unless both chromosomes have the same altered allele, the altered trait generally will not be expressed; it is overridden by the "normal" allele on the second X chromosome. So if a gene predisposing toward male homosexuality were indeed inherited exclusively through the mother it would have to be associated with the X chromosome, which would be a tremendous head start in ultimately isolating and identifying such a gene.

To address this question, Hamer carried out a detailed "pedigree" study of the families of two sets of homosexual male volunteers. Family lineage studies used in genetics normally focus on simple parent-to-offspring inheritance patterns; Hamer decided to explore some of the branches of the trees, as well as their main trunks. One set of seventy-six volunteers was randomly selected, in the sense that the investigators knew nothing of the incidence of homosexuality among family members. The second set was recruited by specifically advertising for brothers who were both homosexual; thirty-eight pairs were selected for the study. Through extensive questioning of the volunteers and of various family members, sixty-nine homosexual relatives were identified from the two sets, and interviews with the individuals identified by family members confirmed their sexual orientation.

The pedigree analysis of the seventy-six randomly selected volunteers showed that, as expected, the highest rate of homosexual orientation—13.5 percent—was in brothers (Table 13.1). Among other male relatives, only two groups had a higher level of homosexual orientation than would be expected by chance: maternal uncles (7.3% gay), and the sons of maternal aunts (7.7%). Among the paired homosexual brothers in the second set, again the only male relatives with a

Table 13.1. *The frequency of homosexual relatives in gay men*

	Group I*	Group II	Controls
Relative			
Brother	13.5	(100)	4.7
Uncle			
maternal	7.3	10.3	1.3
paternal	1.7	1.5	3.2
Cousins			
maternal			
through aunt	7.7	12.9	1.6
through uncle	3.9	0	0.9
paternal			
through aunt	3.6	0	3.2
through uncle	5.4	5.4	3.2

*Group I: Randomly selected volunteers. Group II: Selected gay brothers. Controls: Relatives of lesbian subjects. Figures are percentages of indicated relatives who were gay. Statistically significant values are in bold type. Based on data presented in Hamer et al., 1993, and Hu et al., 1995.

significant level of homosexuality were maternal uncles (10.3%) and the sons of maternal aunts (12.9%). As controls, Hamer looked at the frequency of gay men among the relatives of lesbian subjects used for a separate study, since all evidence suggests that homosexuality in gay men and lesbian women is genetically distinct. These results confirmed in a convincing way the transmission of a predisposition toward male homosexual behavior through maternally inherited genes.

The higher family incidence of homosexual orientation in the families containing two gay brothers suggested such families would be good candidates for tracing the genes involved. The clearly demonstrated maternal inheritance of these genes strongly suggested that they would be found on the X chromosome. Thus DNA samples from the homosexual brothers in forty different families, together with DNA from their mothers, where available, were subjected to a "chromosome scan"; they were typed for twenty-two DNA markers spanning the entire X chromosome. The mother in each case will have two X chromosomes, and the DNA markers (selected in the first place because they are highly allelic in the general population) will be in allelic form at most of the twenty-two marker loci. If there is a gene predisposing to homosexuality on the X chromosome, then gay brothers in the same family should have inherited the same X chromosome from their mother, and therefore the same allelic DNA markers located near that gene. The results showed that markers defining a region of the X chromosome called Xq28, located at one tip of the chromosome (Fig. 13.2), were inherited by 33 (82%) of the forty gay pairs tested. The likelihood that this could happen by chance is small.

In a follow-up study over the next two years, Hamer's group checked their initial results with the Xq28 DNA markers on a second group of thirty-three families having two nontwin gay brothers, and extended their research to include thirty-six families that had two nontwin lesbian daughters. In these studies, they used additional markers more closely spaced around the Xq28 region, in order to define more precisely the chromosomal region influencing male homosexuality. They found that a preponderance of gay brothers (albeit a lower 67%) shared the same Xq28-associated markers; these markers defined a narrower QTL in Xq28 called DXS52 (for the nearest marker). This time they also checked the DNA of heterosexual brothers (where they existed) of the gay subject brothers; only 22

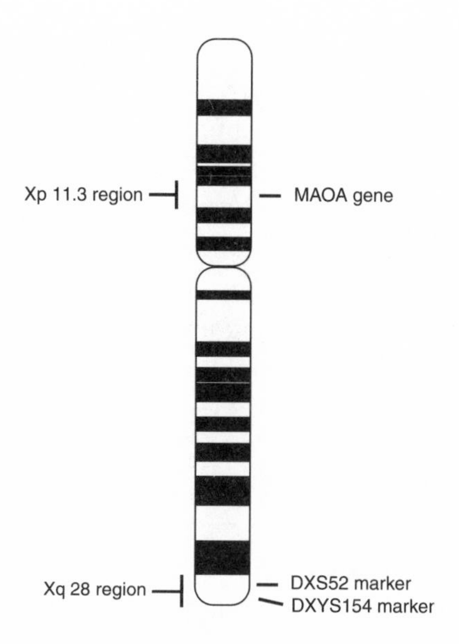

Figure 13.2 The human X chromosome, showing region identified by Hamer's putative DXS52-linked gene. Note also the position of the MAOA gene referred to in Chapter 9.

percent of the nongay brothers had the same DXS52 markers as their gay brothers. When the same analytical methods were applied to lesbian sisters and their heterosexual sisters, no preferential distribution of X chromosome markers of any kind was found. A lesbian was just as likely to share the same DXS52 markers with her heterosexual sister (56% of the time) as with her lesbian sister (58% of the time.) Even females with the appropriate DXS52 markers on both X chromosomes were not necessarily lesbian. Thus it would appear that the X-chromosome gene predisposing to homosexuality in males, while certainly present in females, has no influence on the development of female homosexuality.

Given the polygenic nature of homosexuality, made clear by numerous inheritance studies, we assume the DXS52 marker defines only one of a number of QTL scattered throughout the genome that influence homosexuality. While it is clear that the DXS52 QTL is involved in predisposing some males toward a male sexual preference, it is equally clear that the gene or genes at this QTL are neither necessary nor sufficient for the development of male homosexuality. And it apparently has no effect at all on the development of female homosexuality, which is consistent with previous thinking that male and female homosexuality may be genetically different in at least some respects. In males, the fact that a number of gay brother pairs did not both show preferential inheritance of the DXS52 markers is a clear indication that other genes must be involved. And the 22 percent of the heterosexual brothers in the follow-up study who had the predisposing DXS52 marker but were not gay also indicates that the DXS52-associated gene must require other genetically or environmentally controlled factors to trigger homosexual behavior.

The existence of a gene in the XQ28 region affecting homosexuality has recently been challenged by a Canadian study that failed to find preferential inheritance of markers in this region in male homosexuals. This study did not challenge the notion that homosexuality has a heritable component or that it is linked to the X chromosome, but the researchers disagree about the previous locus. Studies are now looking at how subjects were selected for these and other studies. Hamer purposely selected families where maternal inheritance seemed to be involved. He willingly admits that such individuals may not be representative of all male homosexuals, and that there will likely be many other genes involved in the heritability of this trait. Hamer is now searching the DNA in the region of DXS52 for candidate genes. It is certainly possible that at least one gene, with a significant influence on at least some forms of male homosexuality, may eventually be found at this locus.

Although it will likely be several years before the presence or absence of a DXS52-associated gene will be confirmed or disproved, there is already a good deal of informal speculation about what such a gene—or any other gene associated with homosexuality—might govern. As we saw earlier, the default gender in mammals, including humans, is female. The presence of a gene called *sry*, located on the Y chromosome, is absolutely required for the development of maleness. This gene shuts down female development, and induces the development of male sexual characteristics; in the absence of *sry*, an individual will always be female. Clearly neither primary nor secondary sexual characteristics are affected by Hamer's gene; gay males are indistinguishable in these respects from heterosexual males. However, most of the Y chromosome genes evolved from genes on the X chromosome; it could be that an X chromosome gene evolutionarily related to *sry* affects certain behavioral aspects of female reproductive function, and under certain circumstances can influence male mate selection as well.

It is unlikely that the DXS52-linked gene affects male reproductive hormones such as testosterone; there is no discernible difference in either the timing of appearance or in the levels of these hormones between gay and heterosexual men.* The possibility of different allelic variants in the receptor for testosterone predisposing to homosexual behavior has been examined, but no correlation of a particular testosterone receptor with heterosexuality or homosexuality could be found. The lack of involvement of sex hormones generally in predisposing to homosexual behavior is apparent in the fact that solid indicators of this behavior can be discerned in both males and females well before the increases in these hormones during adolescence. "Gender-atypical" behavior as a forerunner of homosexuality has been noted for over a hundred years. Individuals often display preferences for opposite-gender playmates and play activities, and identify with opposite-gender role models, well before such activities could have a reproductive association. Gender-atypical behavior is apparent in children even before they enter school, and is an accurate predictor of future sexual preference across many different world cultures. We do not know of course that this is the aspect of homosexuality governed by Hamer's X-linked gene, but if it is, we would expect it to act at a very early stage in development, possibly even from birth.

As part of human sexual development, there clearly must be a mechanism in both sexes that favors, in sexually mature adults, selection of a mate of the opposite gender. In the development of male homosexuality, there is not only a strong sexual attraction to males, but a fairly categorical rejection of females as mates. This rejection of the opposite sex is much less evident in nonheterosexual females, as reflected in the higher proportion who are bisexual. There is strong evidence in other mammals that mate selection is not learned behavior, and there is no reason to believe that it is in humans. Whether this mechanism is the same in males and females, we do not know. Evidence from both

---

*There is a condition in human females, called congenital adrenal hyperplasia, in which levels of male hormones are elevated just before and after birth. The incidence of gender-atypical behavior in childhood, and of homosexuality as an adult, is significantly higher in these individuals than in the female population at large.

human and animal studies suggests that this mechanism is in fact genetically acquired. Ultimately, a genetic understanding of both male and female homosexuality will have to include an explanation of this change in the mate selection component of sexuality.

Is there evidence for homosexual behavior in animals other than humans, or is such behavior a unique attribute of the human brain? Many investigators have tried to answer this question over the years. Most scientists agree that homosexual behavior, beyond random same-sex playing and exploration in young animals, does not occur in animal species below the primates (humans, apes, monkeys, lemurs, tarsiers, and marmosets). In some species of primates there is overt behavior that is difficult in many ways to distinguish from human homosexual behavior, in both males and females. However, because we have no way of plumbing the psychological elements that are such an important part of human homosexuality, these interactions in nonhuman primates are usually referred to as same-sex sexual behavior.

In New World primates, which are evolutionarily older, same-sex sexual behavior does not get beyond the juvenile play stage observed in other animal orders. It is in some of the more recently evolved Old World species, such as apes and chimpanzees, that we begin to encounter startlingly human-like, same-sex sexual behaviors. Couples of the same sex often establish and maintain a long-term, stable, exclusive relationship. There may even be competition for same-sex mates, with displays of aggression and apparent jealousy. These couples engage in complex sexual behaviors, including genital-to-genital contact (particularly among females), and mutual genital manipulation.

Nevertheless, there is still disagreement about the extent to which primate behaviors of this sort are equivalent to human homosexuality. A key element, often overlooked, is that there are no well-characterized instances of *preferential* same-sex activity; in every case that has been examined, if given adequate access to a partner of the opposite sex, primates will abandon same-sex for opposite-sex interactions. Often same-sex interactions reflect inadequate access to mates of the opposite sex stemming from dominance structures within a social unit. Interestingly, even within the constrained same-sex sexual behavior paradigm, there is a difference in male versus female activities. Male same-sex behaviors almost always occur in the juvenile and adolescent years, whereas females rarely engage in same-sex behavior during these years; they almost always engage in such activities as adults, even after they have begun breeding with males and bearing young.

These behaviors are clearly distinct from anything found in lower animal orders, and may well represent at least the beginnings of the corresponding neurological and behavioral patterns in humans. Whether or not these behaviors cluster in familial lines in nonhuman primates—whether they are heritable—has never been determined.

The existence of genes predisposing to homosexuality poses an interesting challenge to current evolutionary theory. The conventional view is that when new allelic variants of genes arise through various mutational processes, whether or not these alleles spread from the individuals in which they arise into the general population depends upon how natural selection acts on them. The only way a new allele can be selected for further propagation within a species is if it somehow increases the reproductive success of the individuals that inherit it. And of course, therein lies the rub: How could an allele that diverts individuals from reproductive activities be established and maintained within a species over evolutionary time? Homosexual behavior in humans is at least as old as recorded history, and the existence of something very close to homosexual behavior in many of the higher primates suggests it has probably been present and maintained in humans from close to the time of our origin as a distinct species. How might this have happened? Homosexual men and women produce only about one-fifth as many children as their heterosexual counterparts. Ordinarily this would lead to rapid elimination of the underlying alleles in short order. But that doesn't seem to be happening. Could it be that some of the genetic alleles predisposing to homosexual behavior confer (or conferred in the past) an advantage on human beings that we have yet to discern?

Allelic variants of genes associated with homosexuality are not the only ones whose survival is something of a mystery. It is becoming increasingly clear that many of the mechanisms underlying aging and death are under genetic control. It is difficult to perceive the reproductive advantage of those genes. There are also gene variants that cause hereditary diseases such as cystic fibrosis; as discussed in Appendix II, it has been estimated that 20 percent of the Caucasian population carries a defective allele of this gene. That is an extraordinary frequency for an allele that causes sterility and premature death in those unfortunate enough to inherit two copies. The best guess at present is that variants of genes that interfere with reproduction in the homozygous form may actually confer some unperceived benefit to individuals who carry only a single copy, the so-called "heterozygote advantage." In some cases, such as sickle-cell anemia, we in fact know the advantage tendered by single copies of the errant alleles. But whether, or to what extent, heterozygote advantage could affect selection of genes involved in a quantitative trait like homosexuality is not clear.

Like research into the genetic basis of mental function, research into the genetic basis of human sexual preference—and specifically, a possible genetic explanation of homosexuality—has been criticized on two quite separate grounds. There have been, as expected, valid scientific criticisms of Hamer's study, mostly focusing on the assumptions made by the researchers about the relationship between paired sibling studies and inheritance of complex genetic traits. Hamer's group has responded to these criticisms in exchanges of opinion in scientific journals and at scientific meetings, as is perfectly normal in such cases. Everyone agrees that more data must be collected to ascertain exactly how genes may be involved, and doubtless future research designs may be modified somewhat in response to criticisms that have been made.

But research of this type inevitably evokes strong reactions of a political and social, as well as a scientific, nature. Some elements of society would very much like to believe that homosexuality is a lifestyle choice and thus a moral, rather than a biological, issue. Some would like to prove beyond a reasonable doubt that homosexuality is as genetically driven as any other human behavior, and that the element of choice plays little if any role in becoming homosexual. The lack of significant choice in becoming homosexual could provide a basis for legal protections extended to homosexuals as a class, and many people oppose this.

Yet defining homosexuality as a normal human behavioral variant, with a strong genetic component, will be a double-edged sword even for those who stand to gain the most from such a recognition. Some would like to see the new view of homosexuality used to define it as a disease, in the sense that we now think of alcoholism as a disease. And as discussed in the previous chapter and in Appendix II, one of the greatest fears about human genetics in general is that once we have isolated and deciphered every gene in the human genome, and know exactly what it does, there will be enormous temptation to combine that knowledge with our increasing ability to control reproduction, to favor transmission of those very same genes. It is not at all unimaginable that heterosexual couples with a family history of homosexuality may want to screen the embryos they produce for the underlying genes, and discard embryos they consider at risk for homosexuality. This is an entirely valid fear; the technology is already there, and it is only a matter of time until we know the identity of the genes involved.

Another legitimate worry on the part of some is that the same knowledge that would allow selective elimination of potentially homosexual embryos could be used to determine the probable sexual orientation of adults on the basis of a simple genetic test. This takes us into the much larger arena of concerns about genetic privacy. Health and life insurers might want to use genetic tests to estimate risks of HIV infection, just as they now routinely screen, using standard medical tests, for a range of other health risks prior to issuing insurance policies or determining premiums. Employers hiring into a particular workplace context may decide they want to use genetic information to determine someone's sexual orientation, in the same way they routinely use psychological and personality tests prior to hiring individuals. These are questions we as a society must address, and soon. Hamer and his colleagues were clearly thinking of these questions when they wrote their first paper on the DXS52 locus; here is their closing paragraph:

> Our work represents an early application of molecular linkage methods to a normal variation in human behavior. As the Human Genome Project proceeds, it is likely that many such correlations will be discovered. We believe that it would be fundamentally unethical to use such information to try to assess or alter a person's current or future sexual orientation, either heterosexual or homosexual, or other normal attributes of human behavior. Rather, scientists, educators, policy makers and the public should work together to ensure that such research is used to benefit all members of society.

The overwhelming majority of the scientific community would agree. We must never allow individuals defined by molecular genetics as somehow different from the "norm," on whatever basis, to be put at risk for selective and potentially prejudicial treatment. To the extent that information about genetic constitution is ever used to identify, isolate, or diminish any one segment of society to the advantage of another, all of the potential benefits of our knowledge of human genetics could be lost in a political backlash the likes of which we have not seen since the closing days of eugenics. These benefits are too important to all of us to let this happen. Every segment of society—physicians, scientists, legal experts, ethicists, and most important, common citizens—must make itself aware of the underlying problems and possible solutions, and join in the debate that will decide these issues as we enter into the twenty-first century.

14

Genes, the Environment, and Free Will

Understanding the biological basis of behavioral variability goes to the very heart of how we see ourselves as individual human beings. What makes each of us so different in the way we interact with one another, and with the world around us? There are many ways to approach this question; the one we have taken in this book is to ask whether, and to what extent, the variability we see in behavior among different individuals is likely to be heritable, and whether there is good reason to believe it is correlated at least in part with allelic differences in behavior-associated genes.

We have devoted a good deal of space in this book to discussing behavior in species that are only distantly related to human beings, such as Paramecia, *C. elegans* and Drosophila, as well as more closely related animals such as rats and mice. This helps us to interpret human behavior in a larger biological context. While human behavior may seem very different from animal behavior in many important ways, behavior certainly did not arise in evolution only with human beings.

In fact, we find that even the simplest of living organisms, such as Paramecia, do behave in response to their environment, although their behavior is limited to immediate, reflexive responses. The behavioral alterations we observed in Paramecia—the *pawn*, *fast*, and *pantophobic* mutations—all centered around an inability of the cell to move properly in response to chemical or physical stimulation. These alterations were heritable, and clearly caused by variants of behavior-associated genes.

But already, even at this early stage of the development of behavior, we see something that will prove to be true of behavior-associated genes in all animal species, including humans. Paramecia genes in which allelic variation alters behavior are not dedicated to any particular behavior, but rather direct the daily life and function of the cell. The mutant alleles we studied in Paramecia do not affect the ability of the cell to sense its environment, nor do they affect the physical apparatus used by the cell to move from one place to another—the cilia and the underlying energy conversion mechanisms. The environmental signal was properly detected, but the way the cell processed that signal had been compromised by various genetic alterations. The genes in which behavior-perturbing alleles arose were of two classes: genes involved in signal transduction, and genes encoding or regulating membrane ion channels.

Behavior becomes somewhat more complex in the roundworm *C. elegans,* a simple but multicellular animal in which we see the beginnings of a separate nervous system. Although consisting of only 302 cells, we see all three categories of cells that define all higher animal nervous systems: sensory neurons, interneurons, and motor neurons. The most instructive gene among those we saw in *C. elegans* is the *npr-1* gene. In this case, we saw a clear behavioral phenotype—group versus solitary feeding—markedly influenced by naturally occurring variants of a single gene. The molecular nature of the protein encoded by *npr-1* suggests it is a membrane receptor, probably for a molecule that looks like neuropeptide Y. Presumably, when worms begin to feed, they release a neuropeptide Y-like protein, which is picked up by the *npr-1*-encoded receptor on neighboring cells, which in turn triggers a series of cellular reactions culminating in social feeding behavior. Depending on which allele of the *npr-1* receptor gene worms inherit, they will be guided to engage in a particular type of feeding behavior.

The property affected by *npr-1* is once again simply signal transduction. But now the cells most specifically affected by this allelic alteration are nerve cells.

It might be tempting to conclude that social versus solitary feeding behavior is under control of a single gene. But that would not be an appropriate conclusion. The same solitary feeding behavior would be induced in a group-feeding strain if the gene for the neuropeptide Y-like protein itself were faulty, or if the genes controlling any of the other steps in receiving or processing the neuropeptide Y signal were faulty. If we found such mutants, we might then be tempted to say that social versus solitary feeding was determined solely by alleles of those genes. The *npr-1* gene is important because it is a classic example of how single genes can *contribute* to a given behavior, but are not in and of themselves the sole *determinants* of that behavior, even at the genetic level.

In Drosophila, which has a well-developed, fairly sophisticated nervous system with the beginnings of a rudimentary brain, we again find that behavioral variability is affected by genetic variability, and that the genes involved have a selective effect on nerve cell functions. One of the very first animal behavior genes ever discovered—the *dunce* gene—turns out to be simply a gene for an enzyme involved in regulating a classic signal transductic molecule: cyclic AMP. Cyclic AMP allows enzymes called kinases to alter proteins within cells, and in most cases induces new gene expression. Allelic variants of genes affecting cAMP can have profound effects on the ability of flies to learn and to store information about their environment, and thus directly affect behavior. The *rutabaga* and *amnesiac* genes also regulate cAMP metabolism. Again, none of these are behavior genes per se—they play similar roles in every cell in the body. The *dunce*, *rutabaga,* and *amnesiac* mutations are found in every cell of a fly bearing these mutations, but it is only in cells of the nervous system that we see affects of these mutations on behavior.

While large numbers of behavior-associated genes have been described in simpler organisms, the vast majority of these have been discovered by deliberately inducing mutations. The gene variants produced in this way are in effect simply new alleles of the gene that was mutated, but there is always the concern that the method used to induce the mutation could have altered more than a single gene, or affected the host organism in some undefined way. On the other hand, scanning wild populations of these simpler organisms for naturally occurring behavioral mutants would be extraordinarily difficult and expensive, and is hardly ever done. Nevertheless, a few naturally occurring gene-based behavioral variants have been identified in lower organisms. The *npr-1* variants in *C. elegans* are one example; *forager* in Drosophila is another. Flies with the "rover" allele of *forager* tend to move far afield in the search for food; those with the "sitter" allele show much less tendency to move very far in the search for food. These naturally occurring alleles turn out to be variants of a gene encoding a cGMP*-dependent kinase. Almost certainly there must be a great many naturally occurring behavioral mutations among simpler organisms; they are just technically very difficult to identify. With the limited examples on hand, there do not appear to be any differences between naturally occurring behavioral alleles and behavioral alleles induced by mutation.

The individual genes identified as affecting behavior in simpler organisms can also be found in mammals; *dunce*, for example, or the spectrum of genes regulating ion channels. In mice, in particular, it has been possible to study the effects of individual genes using the gene knockout technique. And in some instances, single genes can exert a powerful influence on mammalian behavior. Prairie voles are monogamous; pairs mate for life, and share equally in raising their young. Mountain voles are just the opposite. They are polygamous, and do not form pair bonds; the male deserts the female immediately after mating, and plays no role in caring for the young. This profound difference in behavior has been traced to two naturally occuring alleles of a single gene, encoding the brain receptor for the neurohormone vasopressin. In fact, the protein receptor encoded by each allele is identical; the difference lies in the control region of the gene, causing it to be expressed in slightly different regions of the brain for the two types of vole. There are likely to be other genes that also contribute to this behavior, but if a genome scan were to be done, the vasopressin receptor gene would likely stand out as an enormous QTL peak on one of the vole chromosomes. Because of its clear role in interpersonal bond-

* Cyclic-GMP (cGMP) is a molecular analog of cAMP, and is involved in similar functions in the cell.

ing, the vasopressin system is now being looked at closely in children afflicted with autism.

The NMDA gene, alleles of which so profoundly affect learning in mice, could be cited as another example. Still, in more complex animals single genes more often than not have only a minor effect on behavior, so scientists have shifted their emphasis from the study of individual genes in mammalian behavior to the study of gene complexes. Undoubtedly, behavior in simpler organisms is also multigenic, but single genes more often have a major impact on a given behavior. The validity of the techniques for genome-wide searches to identify behavior-associated gene complexes (QTL), which holds such great promise for human behavioral genetics, has been amply established in rats and mice. Because the genetics and biochemistry of rodents and humans are so similar, genome scans of rodent species will be a valuable adjunct to our exploration and interpretation of behavioral QTL in the human genome.

So there are at least two important lessons learned from studying the genetic bases of behavior in animals that will likely apply to behavior in humans as well. First, there are unlikely to be "behavior genes" per se; the genes that influence behavior appear to be the same genes that direct the day-to-day operation of cells in general, but nerve cells in particular. And second, even relatively simple behaviors in humans will almost certainly involve the coordinated interaction of many genes, both with each other and, as we discuss below, with the environment.

Genetic Contributions to Human Behavior

Our understanding of the contribution of genetic polymorphisms to human behavioral variability lags far behind our understanding of the role genes play in animal behavior, for the simple reason that we are greatly constrained by the types of experiments we can carry out in humans. The vast majority of behavioral genes in animals were discovered through induced mutation, which is clearly not acceptable in humans. In animals, once a mutation is suspected to have behavioral consequences, its genetic basis can be confirmed through controlled breeding studies, in which the altered phenotypic trait is followed through multiple successive generations. Again, this is not something we can or should do in humans. Once we are convinced by inheritance patterns that the change is indeed genetic, we can begin a search for the gene itself at the DNA level. Once we find the gene, we can then move various allelic forms of it between animals. In humans, we are restricted to observing the unregulated transmission of phenotypic traits between generations. Because of the much longer generation times, and the random nature of human reproductive activities, this approach is not particularly efficient scientifically, but it is the only approach we have to analyzing gene involvement in human behavioral variation.

Serious investigations of familial transmission of human behavioral variability are only a few decades old. Many of these studies were, in the earlier years, often little more than anecdotal discussions of apparent inheritance patterns in small numbers of families. These have evolved into sophisticated, statistically sound analyses of data obtained from large-scale family studies, including both biological and adopted siblings and analyses of monozygotic and dizygotic twins reared together or apart. The earlier studies were rightly criticized on a number of points relating to sampling size and methods, as well as statistical interpretations of the data. But these criticisms have been addressed in more recent studies, and such studies now provide compelling evidence that a significant portion of the observed variability in human behavior is indeed heritable.

Although following the passage of genetically influenced traits from generation to generation can establish heritability of a given trait in human beings, it does not allow us to directly identify the underlying genes. In the past dozen years or so, however, new developments in the fields of genetics and molecular biology have made possible the beginning of direct, systematic searches of human DNA itself for transmissible genetic factors underlying human behavioral variability. These approaches have been greatly facilitated by the technology emerging from the Human Genome Project, and as this project moves toward its conclusion in the year 2003 or so, we will see even more tools and resources emerging that will aid in our search for genes affecting behavior.

Genome scans, in particular, will profit greatly from information emerging from the genome project. This approach, more than any other, has made abundantly clear that virtually all behaviors are genetically complex, involving the interaction of many different genes. We are just now seeing the first data emerging from genome scans: Numerous QTL have been identified as housing genes whose alleles affect alcoholism, eating disorders, substance abuse, and other complex human behaviors. We will likely know the exact identity of many of the genes at these QTL in the next ten to twenty years. But identifying the genes involved will be just the beginning. Sorting out how all the allelic variants of these genes interact with one another and with the environment to influence particular behaviors will likely occupy scientists for decades to come.

But while inheritance studies and QTL scans are highly consistent with a role of genetic variation in human behavioral variability, it must be admitted that at present we have only limited and indirect insights into the nature of the genes involved. One category of genes that will almost certainly be involved in human behavioral variability is that encoding neurotransmitters and their receptors. The ability of neurotransmitters to affect behavior is beyond question. In humans, variations in serotonin levels are associated with behaviors ranging from aggression to depression, and allelic variants of genes for both neurotransmitter-synthesizing enzymes and neurotransmitter receptors have been found in humans. Studies in animals have made it clear that variations in neurotransmitter pathway genes are not just associated with variability in behavior, but actually cause that variability. There is no reason to believe this would be different in humans.

Neurotransmitter genes may be as close as we can come to behavior-specific genes. They are relatively specific to the nervous system, which is concerned with little else than behavior. What we do not find, however, is a one-to-one, dedicated correspondence between a given neurotransmitter genotype and a given behavioral phenotype. What we find, rather, is that neurotransmitter systems, particularly in humans, cut across wide ranges of behaviors, affecting them in ways that would be difficult to predict in advance.

We see over and over again that the same genes in which natural or induced allelic variation cause behavioral alterations in other species, whether controlling ion channels, signal transduction, or neurotransmitter function, are also present in human beings, and govern similar cellular functions. This raises an important question: Is there any reason to expect that naturally occurring variants of these genes in humans would not cause variability in human behavior as well? Can we imagine that the connection between genetic and behavioral variability we see in animals is suddenly lost in evolution with the emergence of human beings? Are we, as a species, fundamentally different from other species in the mechanisms governing our behavior?

Human beings are clearly different from animals in that humans have developed something called mind, and from mind has come culture. A great deal of what we do—a great deal of our behavior—can originate in the mind, with relatively little immediate input from the physical environment. The environment, when we respond to it, may itself be a product of mind—in many cases our behavioral responses are directed to the intellectual and cultural constructs we have surrounded ourselves with, as much as to our ecological environment. The question thus becomes, is the resulting behavior under control of genes in the same way as our behavior in response to the physical environment? Will variability in those genes be associated with behavioral variability in the same way we see in other animals? Does the acquisition of language and culture demand that the underlying basis of human behavior be substantially different from the basis of behavior in any other species?

Some have argued that culture now has a life of its own, even that it evolves independently of our genes, and controls human behavior in ways that simply do not apply to animals. That may well be true, but intellectual and cultural information from our environment is absorbed and processed by individuals in the same way as information about the physical environment. All human behaviors—even behaviors in response to constructs of mind and culture—are mediated by the nervous system, by our brains. The same neural pathways that process the meaning of songs and novels, political arguments and religious beliefs, also process our responses to predators, food, the weather, and potential mates. What we do with our nervous systems in terms of behavior may seem much more complex than what a fruit fly or a worm does with its nervous system, but the *way* we do it has changed very little over evolutionary time. The molecular mechanisms, and the genes underlying those mechanisms, involved in operating a fly's neurons are surprisingly similar to those governing a human neuron. The difference lies in the extent and complexity of neuronal interconnections;

Table 14.1. *Genes, genomes and nerve cells. The proportionally greater increase in DNA versus numbers of genes during evolution reflects the increasing accumulation of noncoding ("junk") DNA.*

Species	Genome Size*	Estimated Number of Genes	Nerve Cells
Bacterium	1-2	1,500–2,000	0
Yeast	13–15	6,000–7,000	0
C. elegans	100	17,000	302
Drosophila	170	14,000	250,000
Humans	3,000	70,000	100,000,000,000

*Millions of pairs of nucleotides

the mechanisms for both generating and operating those interconnections differ very little. Even the most complex and sophisticated operations of the human mind are mediated by the same underlying neural mechanisms that draw a fly to a piece of fruit, or cause a mouse to hover uncertainly in an open-field test.

At one time it may have seemed that the genetics of human behavior would prove to be much more complex than the genetics of animal behavior because humans have so many more genes than simpler organisms. But this turns out not to be true. Humans, with tens of billions of brain cells, are enormously larger and more complex at every level than the roundworm *C. elegans*, with its 302 nerve cells. Humans are estimated to have about 70,000 genes; yet we know from direct sequencing of the *C. elegans* genome that these tiny creatures already have just under 20,000 genes (Table 14.1). Moreover, humans often have more copies of a given gene in their genomes than simpler organisms. Humans have at least seven forms of the gene for phosphodiesterase, whereas Drosophila has only three; humans have over a dozen different genes for serotonin receptors, whereas *C. elegans* has only two. In fact, although the *number* of genes in humans is more than three times as great as in *C. elegans*, the number of *kinds* of genes may not be hugely different. So genetic complexity per se seems an unlikely explanation of human behavioral complexity.

The Impact of Environment on Behavioral Variability

One of the major problems with the direct application of what we have learned about behavioral genetics in animals to humans is that only a handful of studies in lower organisms have specifically addressed the role of the environment in determining behavioral phenotypes. Most animal studies focus on the genetic aspects of behavior, and great pains are usually taken to remove environmental variables that might affect behavior. This is a necessary step if we are to identify clearly the genetic contributions to individual behavioral traits. But the same evidence that tells us about the contribution of genes to human behavioral variability also tells us that the environment accounts on average for at least half of that same variability.

Among the few animal studies directly aimed at exploring environmental variables have been those showing that almost any form of stimulation, even simple handling or exposure to strange smells and sounds during the early, preweaning stages of life correlates with decreased fearfulness, and an increased willingness to explore new experiences and sensations, later in life. These experiments have been repeated with many different animal species, and appear to be generally true. Experiments with monkey infants have demonstrated the same thing, using an opposite experimental approach. When monkeys are placed immediately after birth into total isolation with only a milk-dispensing, cloth doll as a surrogate mother, their personal and social behavioral development is drastically stunted. If introduced into a normal environment after three months of such isolation, they display extreme fear and social dysfunction, but most eventually recover. If isolation is extended to six months or more, they develop something akin to severe autism, and may never recover. They are unable to function in ways that would be necessary to survive on their own; they do not even engage in a behavior as fundamental as sex.

In animals, the impact of environmental experience on the brain itself is abundantly clear. Studies beginning in the 1950s have recorded the physical changes that environmental experience and learning tasks impress on the nervous system. Visual or olfactory stimulation, as well as motor activity, can increase the size of the corresponding regions of the brain by 20 percent or more. These increases result from increased numbers of neurons, as well as increased numbers of dendrites and synaptic connections serving individual neurons. There are also increased numbers of blood capillaries bringing oxygen and nutrients to support the increased activity of brain cells. While the connection between these structural changes and environmental experience is indisputable, it is difficult to interpret them qualitatively, at the level of specifically altered behavioral responses.

Systematic studies of the influence of environment on behavior in humans are even more problematic. The same ethical restrictions that prevent us from direct manipulation of the human genome, to gauge the impact of genes on behavior, also prevent us from deliberate manipulation of the environment for the same end. We are largely restricted to observing how individuals exposed to different environments at different stages in their lives behave, and trying to draw meaningful conclusions about what we observe. Unfortunately, a good deal of research along these lines has tended to be anecdotal, often based on limited numbers of observations, and evaluated along confusing and often contradictory paradigmatic lines.

As with studies of genetic influences on behavioral variability in humans, most of the quantitative information we have on the role of environment comes from studies of twins reared together and apart, and from adoption studies. These studies are by necessity observational; the influence of the environment is deduced mostly as anything that cannot be accounted for by inheritance. Statistical interpretations of these data raise interesting speculations, but these cannot be tested by direct experimental manipulation of individual human beings. It is almost certainly true that the environment has a major impact on human behavior, but support for this notion is largely inferential; there is admittedly very little hard data.

On the other hand, in humans, as in animals, there is evidence that environmental experience affects basic brain structure. In general, people who have had a great deal of cortical stimulation during their lifetimes show more extensive dendritic networks in the cerebral cortex at autopsy. Athletes have more highly developed neuronal networks in those regions of the brain involved with muscular coordination. More impressive are some of the studies involving real-time imaging of the human brain using techniques such as magnetic resonance imaging and positron-emission tomography (PET) scans during learning situations involving both verbal and motor skills. PET scans, which monitor metabolic rates inside cells, have also been used to monitor brain cell changes before and after treatment for behavioral disorders. In a recent UCLA study, it was shown that equivalent, stable metabolic changes were induced within cells of the midbrain region of patients with obsessive-compulsive disorder, whether they were treated with a serotonin transporter inhibitor or by behavior modification therapy. This provides a valuable adjunct to studies that examine physical changes such as synaptic number and strength induced by behavioral alterations. It also highlights the power of environmental experience to alter neurological structures.

As pointed out by Richard Lewontin and his colleagues in *Not in Our Genes: Biology, Ideology, and Human Nature* (1984) we must not view genotype and environment as two separate and unrelated forces, acting in opposing or even disconnected ways to shape the final phenotypes that emerge in each human being. Genetic variation can affect brain structure itself during embryonic development, because genes guide developmental processes. Genes will also influence the nature of ion channels, signal processing within nerve cells, and the neurotransmitter systems neurons use to communicate with each other. The signals that are processed are provided by the environment, independent of brain structure and function. These signals can also effect changes in the brain; they can alter intracellular processing pathways, and can induce a remodeling of brain architecture and synaptic connections.

An individual's genotype will markedly affect how he or she sees the environment in the first place. The very same environment that seems hostile and threatening to one person may seem familiar and comfortable to someone else. And our experience in testing our genotypes against the environment will have a lasting impact on our subsequent behavior. We must also bear in mind that individuals, guided by their individual genotypes, manipulate their environments in different ways. Remember, for example, how the fraternal twins in Chapter 1 manipulated those around them differently. Peer group selection, the choosing of friends we feel most comfortable with, creates a powerful environmental influence, but is in turn largely a reflection of genotype. So when we speak of relative percentages of genetic versus environmental influences, we must always remember that these two forces interact with, rather than oppose, one another in the shaping of a human being. Trying to play number games in the nature-versus-nurture debate to see which side "wins" is not only an unproductive use of time, but will almost certainly lead to incorrect conclusions.

Genes, the Environment, and Free Will

Discussions of the relative roles of genes and environment in determining who we are ultimately impinge on the question of free will and personal responsibility. Free will implies freedom from any sort of determinism, whether historical, genetic, or divine, in making choices. All legal and moral systems assume that individuals are free to choose among alternative courses of behavior; individual responsibility has no meaning in the absence of unimpeded choice. But does such freedom really exist? If it does, what is its biological basis?

The study of human behavior evolved within the framework of two rather different points of view, neither of which is held in its strict form by contemporary behaviorists. Biological determinism posits that our every reaction to what happens around us can be predicted from what is written into our genomes at the moment of our conception. Once we have identified all of the genes affecting a given behavior, and once we understand how all of the various alleles of those genes affect that behavior, then we should be able to predict what an individual will do in terms of that behavior in response to a defined set of environmental conditions. The environmental determinist position is that a newborn human being is essentially a blank slate, and that what an individual becomes as an adult is predictable in terms of the cumulative environmental and cultural experiences to which he or she has been exposed. Behavior in response to a defined set of environmental conditions should be entirely predictable from the sequence and sum of previous life experiences.

It is now clear that human behavior is best explained by a combination of the biological and environmental positions, with genes and previous experience contributing roughly equally to the variability we observe in the way humans behave. But neither of these behavioral components provides any insight into free will. In fact, each argues strongly against its existence. If our every behavior can be predicted from what is written into our genomes before we are born, what does that say about our freedom to choose, or about individual responsibility for the choices we make? But those who fear tyranny of the genes can hardly take comfort from the view that we come into the world as some sort of blank template, and that what we become, as cognizant adults, is simply the totality of our previous experiences. That would render our behavior no less predictable—and our freedom to choose and act no less constrained—than if we were simply the sum of our genes.

But the "modern synthesis"—combining genes and environment as an explanation of behavior—leaves us with the same problem with respect to free will. Combining two forms of determinism gives little insight into the origin and meaning of choice in human behavior. Given that free will is an expression of human behavior, and that behavior is rooted in the nervous system, where in this dense complex of interacting nerve cells is free will to be found? Can free will have any meaning in biological terms? Free will is traditionally considered to be a uniquely human attribute. But animals clearly do make choices: Even a *C. elegans* worm, faced with two equal patches of bacteria, decides to eat one, risking the possibility of losing the other. Is this a random choice? Any time there is more than one possible way to react—to behave—an animal is forced to make a choice. Does making such a choice involve free will? If not, what is it we believe is unique about human decision-making processes?

The standard answer would be that human beings can distinguish right from wrong. "Right" and "wrong," in this sense, are cultural constructs, with at best a remote basis in the biological realities shaping the evolution of behavior. But as discussed earlier, the neurological machinery for processing cultural information is the same as that used to process information about the physical environment. It follows that the same allelic variation in genes, the same differences in environmental history, that affect other behavioral decisions will affect our distinctions between right and wrong. But if both genes and previous experience are deterministic, again, where does free will originate?

Insight into this question may well come from a seemingly unlikely source: mathematics, and the nature of prediction. Predictability is the essence of science. A phenomenon is observed; various hypotheses are put forward to explain it; those hypotheses are then tested by their ability to predict what will happen under a given set of circumstances. Competing hypotheses are gradually winnowed down until a unifying theory, with maximal predictive power, emerges. But always it is the ability to predict that determines how long a theory lasts; when a theory is unable to predict an outcome, the theory is modified or, in some cases, abandoned altogether.

The physical theories derived to explain the universe have tremendous predictive powers. They allow us to predict tides and eclipses, the motions of planets and galaxies, or the rate of decay of a radioactive substance. But in a surprisingly large number of cases, even when we understand in great detail all of the physical forces acting in a given situation, we are unable to predict what will happen: the flight path of a balloon propelled by escaping air, for example, or next week's weather. One characteristic shared by most of these seemingly unpredictable systems is that they can be greatly perturbed by slight changes in initial conditions. For example, imagine a small boulder balanced at the crest of a steep hill. We give it a nudge, and it begins tumbling down the hillside. Try to predict its pathway through its descent. Every twig it encounters on the way down, every collision with a bush or stone, sets up a new sequence of future interactions as the boulder descends. And the slightest change in initial conditions—whether and at what angle it rolls over a small pebble in the first inch of its pathway—can have highly unpredictable consequences for the remainder of its journey.

For a long time scientists assumed that the inability to predict behavior in such situations was just a computing problem; given enough detailed information about all of the conditions on the hill, and about the conditions of the initial push that started the boulder on its course, and a big enough computer, they could predict where the boulder would end up. They couldn't. At each step along the descent, each encounter sets up a whole range of possibilities for the next encounter. The final calculation is not just the linear sum of all these encounters, it is the nonlinear *multiplication* of the possibilities generated by each encounter. Within a meter or so of the beginning, the possibilities will already have exceeded what a computer could calculate, or the human brain could comprehend. As if that weren't enough, we also know from quantum mechanics that the very initial conditions of any event—like nudging a boulder from its resting place—cannot itself ever be known with absolute precision; measuring position or momentum at a fine enough level alters either position or momentum. Systems in which slight changes in initial conditions rapidly generate incalculable outcomes are referred to by mathematicians as chaotic.

The brain certainly offers plenty of opportunity for generation of nonlinear, chaotic behavior. Individual nerve cells, as we have seen, are themselves incredibly complex. Information is brought to each cell through numerous extensively branched dendrites; a single nerve cell may receive information from a thousand or more other neurons, each of which was itself impacted by a hundred or a thousand inputs. The possible combinations of dendrites feeding into a neuron at any given moment, plus the range of possible intensities and frequencies of those signals, already sets up an enormous range of possible initial conditions. The electrical potential used to forward information through the cell's single (but again extensively branched) axon is controlled by hundreds of independently operating ion channels, which regulate the flow of sodium, potassium, and calcium ions in and out of the cell. The number of possible combinations of open and closed channels adds another huge factor to determining the state of the cell. Finally, the tens or hundreds of thousands of cells that make up a particular nerve cell tract in the brain may interact with each other in feedback loops, as well as with other cells in regions of the brain targeted by the tract. It is not hard to imagine that the slightest variation of initial conditions in just one neuron—the particular selection of dendrites activated at any given moment, the number and state of active ion channels—coupled with the enormous range of possible interactions among nerve cells within a tract, could produce alterations whose impact on behavior would be as unpredictable as the pathway of a hypothetical boulder down a mountainside.

Why would we want to have programs embedded in crucial organs such as the brain that generate unpredictability? One advantage could be diversity in the ability to respond to the environment. We already have one source of diversity in the presence of multiple gene forms (alleles) within the species. Some alleles of a given gene will work best under one set of conditions; other conditions may favor a different allele. Human beings on the whole are thus equipped to face a wide range of internal and external conditions. Within an individual, systems such as the immune system can even scramble certain genes internally to create the incredible diversity needed to deal with the wide range of pathogens that invade our bodies and cause disease. But that is a rare exception; inherited genes are in general not altered during an individual's lifetime. Chaos does not alter genes, but it could create new—and unpredictable—possibilities in the pathways specified by genes and altered by experience. While we have no direct evidence one way or the other, it is certainly conceivable that nonlinear amplification of small fluctuations in initial conditions within an individual nerve cell or nerve tract could result in the creation of unpredictable patterns of signal processing, resulting in behaviors that are neither embedded in DNA nor related to previous experience. That fits very nicely the definition of chaos.

We are just at the beginning of our attempts to understand the role of chaotic processes in human behavior, but already a number of possibilities suggest themselves. Creativity has always puzzled us. We sometimes make strange mental leaps, arriving at conclusions for which nothing in our previous experience seemingly could have prepared us, but which allow us to see things in an entirely new way. And what about decision making? How many times have we thought back on a particular situation and wondered why we acted as we did? The nervous system is at one level deterministic, but we cannot always predict what it will do. Chaos provides us with a rational basis for expecting that some outcomes of behavior may be too complex to be predicted, that behaviors can emerge that seemingly have no rational basis, either in genotype or in terms of previous experience. On the other hand, the indeterminancy contributed by chaos need not be infinite; the number of different pathways a boulder follows down the hillside may be too large to calculate, but it always ends up at the bottom of the hill.

And that brings us back to free will. We speak of free will as though it is something we control. But is that necessarily true? Our decisions to act in a certain way in a given situation arise entirely within us. But the very definition of chaotic behavior suggests that it would operate outside of human consciousness and memory. So if chaos is a factor in generating human behavior, then it may be that what we are calling free will is simply a way of accounting for a certain level of longed-for indeterminacy in our behavior, of trying to fit it into a pattern that we can understand—and think we can control.

What does this view of free will say about individual responsibility? Our genes and our past experiences, as best we understand their role in human behavior, are highly deterministic. If that's all there were, there would be little or no free will; we could well plead for our every action that we are the helpless victim of one or the other form of determinacy. Such explanations of criminal behavior are increasingly commonplace in our courts. But the indeterminacy of chaos that frees us from both our genes and our past history also forces us to accept responsibility for how we act. Chaos may indeed force us to experience things not scripted in either genes or experience, but we have an extraordinary power to learn. We can see and understand fully how our behavior affects our own lives, and the lives of those around us. Perhaps therein lies the definition of moral choice, and thus the nature of free will; the ability to choose among personal and social possibilities dictated by neither genes nor experience.

The nature/nurture debate will not likely disappear in the near future. We have tried to make the case here that, contrary to the wishes of some, genes do play a very important role in human behavior; we pretend otherwise at our peril. But the truth lying at the end of this debate may very well be that we that we will never be able completely to untangle genetics from environment, or to untangle either of these from a certain built-in randomness in our behavior that sets up the possibility of free will. The working genes we as a species carry with us today have been repeatedly pounded and shaped by the environment; our DNA is itself a palimpsest on which previous experience with the environment has been repeatedly written and erased. Even the "nonsense" DNA that we so laboriously carry around with us from generation to generation is rife with the imprints of fossil genes reflecting previous environmental experiences. The DNA we each possess is just one "freeze-frame" in the genetic history of our species, and none of us is heir to the environmental experience of humanity as a whole. Each of us must struggle to maximize the genetic hand we have been dealt, played in the context of the environment into which we are born, against a certain level of indeterminacy we must somehow learn to bring under control. It is this struggle that defines us, and makes us human.

Appendix I

Finding and Identifying Genes

Most of what we know about human genetics has been learned by following the transmission of phenotypically observable traits from one generation to the next. We present a good deal of indirect evidence in this book that a significant portion of the variability we see in these traits among individuals is heritable, and thus can be attributed to corresponding variations—alleles—in the genes underlying the traits. Since allelic variations are ultimately defined by slight variations in the exact sequence of nucleotides making up a gene, then if we are to understand fully how different alleles contribute to behavioral variability, we need to be able to isolate the genes involved and study their allelic patterns. But how do we go about associating a given trait with a given gene (or genes) in the first place? And how do we find a previously unidentified gene buried among billions of nucleotides spread out among twenty-three different chromosomes?

This problem has been approached in a number of ways since geneticists first began chasing genes at the beginning of the twentieth century. Classical "transmission genetics" was based on the induction of phenotypically observable mutations that could be followed from generation to generation. This allowed the identification of genes involved in a given trait, but provided no clues as to whether other genes might also be involved in the trait, or to the nature of the gene itself. The first gene—from a virus—was isolated only in 1975; the first human gene, encoding the blood protein called β-globin, was not isolated until 1977. Over the next twenty years, only a couple of thousand more human genes had been found, and those were mostly the easy ones.

The techniques used to trace a gene to its chromosomal location, and then to isolate, clone, and sequence that gene, were until recently very arduous. But technological breakthroughs in the past few years, stemming mostly from the Human Genome Project, have greatly simplified this entire procedure, and it is now expected that virtually every gene in the human genome will have been isolated and sequenced by the year 2003.* Unfortunately, we won't know what the vast majority of those genes do; we'll simply have their sequence. But the same technology that led to gene isolation for sequencing can be used in conjunction with existing genetic techniques to determine the function of genes as well. This combined approach, which is already under way with the first genome scans, will be a major step forward in the field of human behavioral genetics.

The most important breakthrough in the ability to find genes and follow them from generation to generation has been the development of something called "DNA markers." DNA markers are small regions of defined nucleotide sequence that occur at a specific place, on a specific chromosome, in the genome of a given organism. If we have a way of finding a given marker, then we will always know exactly where we are in the genome—on which chromosome, and at which tiny region of that chromosome. And if we can associate the inheritance of a given trait with the inheritance of a specific marker, we will know that a gene involved in that trait must lie very close to the marker.

There are two important requirements for a DNA marker. It must be unique—it must occur at only one place in the entire genome—and it must be polymorphic—it must have at least two alleles in the population we wish to study, but preferably more. A marker can be any defined stretch of DNA. It can itself be a gene. Any of the several thousand human genes whose precise location is already known serve as markers for the chromosomes on which they are found. But more recently, advantage has been taken of the special properties of nonsense DNA—the noncoding DNA that makes up the bulk of the human genome—to define a new battery of DNA markers that have nothing whatever to do with genes. One example of this type of marker is the

* The Human Genome Project, and the development of the technology that made it possible, are described in W. R. Clark, *The New Healers: The Promise and Problems of Molecular Medicine in the Twenty-first Century* (New York: Oxford University Press, 1997).

so-called "short tandem repeat," or STR, as it is commonly known.

It turns out that the nucleotide sequences for nonsense DNA, which is scattered more or less evenly throughout the genome, tend to vary a great deal among individuals; there is apparently no selective pressure operating to prevent nucleotide changes and what, in functional genes, might be considered mutations in noncoding DNA. Moreover, a significant portion of this DNA is characterized by large numbers of end-to-end repeats of small numbers of nucleotide motifs—for example, $(ACAG)_n$, where *n* could be anywhere between 50 and 1,000. Clusters of these STRs are scattered throughout the genome. There is an almost endless variety of them, based on nucleotide composition, as well as motif length, and many of them are highly polymorphic in the human population. The fascinating thing about STRs is that their polymorphism lies not in their nucleotide composition, but in the *number* of back-to-back repeats of a given motif an individual may have. For example, one individual may have twenty-six repeats of an ATGT motif at a defined position on chromosome 12; another individual might have $(ATGT)_{31}$ at this same site. Distinguishing the number of motif repeats an individual has at a given locus is relatively easy. So, STRs satisfy the requirements for DNA markers—they are unique for a given genomic site, and they are polymorphic.

DNA markers have now been established for the full length of every human chromosome, and for most species of important laboratory research animals. This "genome mapping" has been a key accomplishment of the Human Genome Project so far. The DNA markers are spaced closely enough together that each of them defines what is, from a molecular biologist's point of view, a manageable segment of DNA. To see just what manageable means, in this context, let's look at an important use of DNA markers—to locate a gene for a given trait. Let's imagine a hypothetical species of lab animals with only two chromosomes, and let's further imagine that each chromosome is defined by just two different DNA markers. (In reality, there would be hundreds.) Some of the members of this species have bright red fur, and it passes from generation to generation as a dominant trait—up to half of the offspring of a red-furred parent will have red fur. The rest of the members of the species all have brown fur. We would like to know where the gene controlling the trait for red fur exists in the genome, so we can isolate it for study.

We begin by breeding a red-furred male with a brown-furred female (Fig. A1.1). Before we do this, though, we first look to see which alleles each individual has of the DNA markers we will use. The red-furred male has the *a* and *c* alleles of marker 1 on chromosome 1, and the *e* and *c* alleles of marker 2. On chromosome 2, the male has the *d* and *b* alleles for marker 7, and the *f* and *a* alleles of marker 9. We can use these alleles to follow the individual chromosomes into the next generation of offspring produced by this individual. Similarly, the female's chromosomes are marked by the *f* and *d* alleles for marker 1,

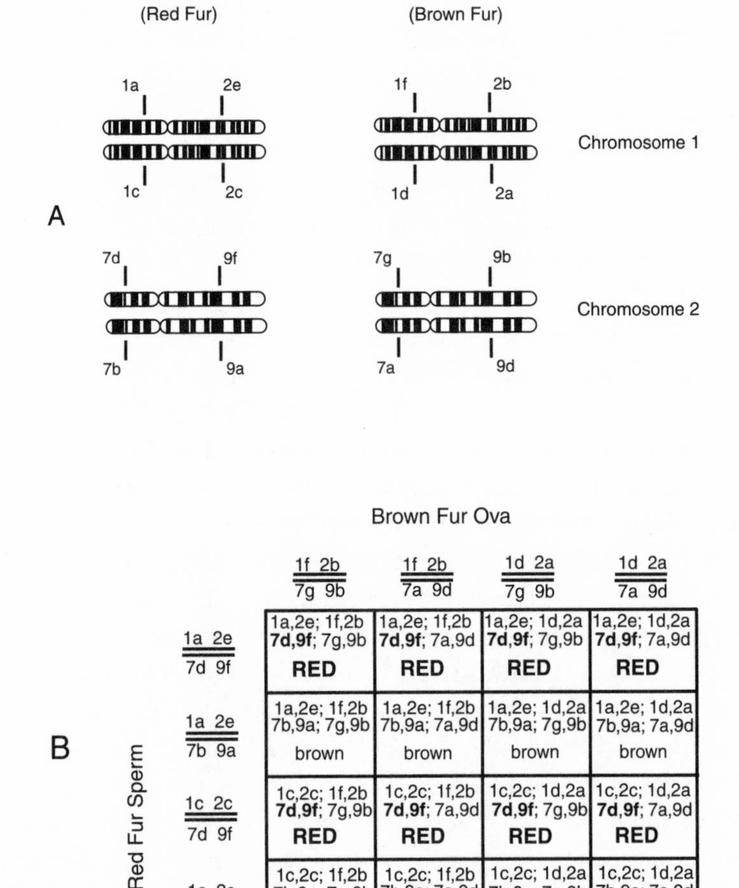

Figure A1.1 Following DNA markers in a breeding experiment. The markers shown on the chromosomes show up in various offspring, represented by the boxes in B.

the *b* and *a* alleles of marker 2, the *g* and *a* alleles for marker 7, and the *b* and *d* alleles of marker 9.

When animals breed, they reduce the number of copies of chromosomes in their sperm or ova from two copies to one, in a special process called meiosis, and the different chromosome half-sets assort randomly into different sperm or ova. Thus, as shown in Figure A1.1B, the brown-furred female can produce four different kinds of ova by randomly distributing her two different copies of each chromosome. (Individual chromosomes are here represented as single lines bearing the marker numbers.) The same is true for the red-furred male. When a sperm and ovum unite, they combine their chromosomes, and the chromosome copy number per cell is brought back up to two in the newly formed embryo. In the figure, the different possible combinations of chromosomes in the embryo, identified by their marker alleles, are shown in the central boxes. Four different sperm and four different ova, when recombined, produce sixteen different kinds of embryos; these can be absolutely distinguished from one another by testing for the presence of the markers. In other words, the exact genotype of each offspring, in terms of its inherited chromosomes, can be determined.

It is then simply a matter of matching inherited genotype with inherited phenotype. Those offspring with red or brown fur are noted in the boxes just below the genotype. We examine the genotype, and look for DNA markers that always travel along with the phenotype. It is clear that red fur appears only in those offspring bearing the chromosome distinguished by the 7d and 9f markers. Therefore we can conclude that a gene responsible for red hair lies on chromosome 2. The markers themselves in this case are not genes; they do not encode a protein associated with red hair. They simply allow us to identify the chromosome on which the gene occurs.

We can refine our search further, and find out which of the two markers on chromosome 2 the real gene lies closer to. When animals produce sperm or ova through the process of meiosis, the original diploid pair of chromosomes often exchange homologous portions with one another (Fig. A1.2) before separating. This is a mechanism for further shuffling of genetic information between generations. After this "crossing-over" event, as it is called, the individual recombined chromosomes are apportioned out into two different daughter cells,

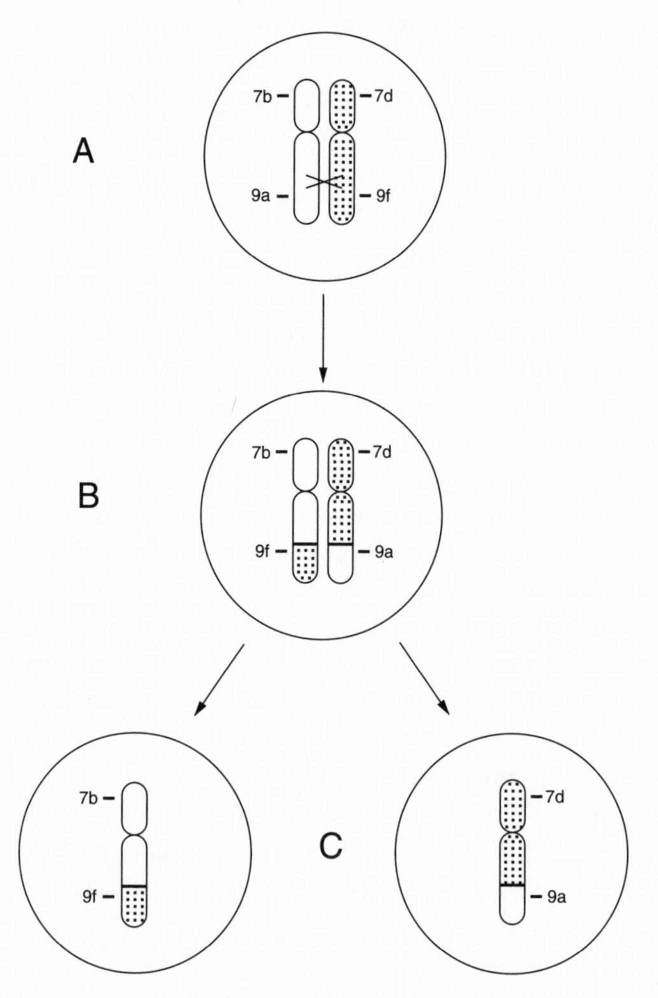

Figure A1.2 Recombination between members of a chromosome pair (A, B) during generation of haploid germ cells (C).

which become the final sperm or ova. For simplicity's sake, only one chromosome pair is shown here, but crossing-over could occur within any or all of the chromosome pairs contributing to newly formed sperm or ova.

If the crossing-over event occurs between two markers, the result will be formation of new marker combinations, as shown in Figure A1.2C. We can then ask which of the markers the trait we are following is most closely associated with. For example, markers 7d and 9a were originally on the same parental copy of chromosome 2. Let's imagine that after a crossing-over event, these two markers ended up on different versions of a recombined chromosome 2, and that the trait of red fur followed the 7d marker. We can now say that the gene affecting red fur color is between the crossing-over point and the end of the chromosome defined by the 7d marker; it "travels with" the fragment of chromosome 2 bearing marker 7. We can thus eliminate the other fragment of chromosome 2 from future considerations in our hunt for this gene. By using ever more closely spaced markers, coupled with crossing-over events during formation of sperm and ova, we can further narrow the region of chromosome 2 in which the gene of interest must lie.

Various versions of this approach underlie virtually all contemporary schemes for finding genes. Since humans produce fewer offspring than most animals, it is necessary to trace marker/phenotype associations in large numbers of families, hopefully through several generations. In genome scans, both parents are described as thoroughly as possible with respect to a given behavioral phenotype, and their DNA is defined by a number of markers associated with each chromosome. The offspring are similarly assessed for the phenotype, and then are tested to see which markers they inherited from their parents. Where possible, grandchildren or grandparents are analyzed in the same way. In a first pass, the results usually just tell researchers which chromosomes are likely to contribute to the trait, and in approximately which regions of those chromosomes QTL are likely to be found (see, e.g., Figs. 10.2 and 11.1). In subsequent scans, researchers will ignore chromosomes showing no activity in the first scan, and will focus on more closely spaced markers in the chromosomal regions testing positive.

A problem limiting this approach is that the likelihood of crossovers between markers decreases as the distance between them narrows, so it is difficult to home in closely on very small chromosomal regions. Nevertheless, with large enough studies, coordinating efforts in various research centers around the world, it can be expected that smaller and smaller regions housing genes of interest will be identified. In the "old days" of molecular biology (five years ago), researchers at this point would have begun the extremely laborious task of sequencing the DNA around the markers until they found candidate genes. Now, researchers can simply scan through the nucleotide sequences already found for the chromosomal region they are interested in through the

Human Genome Project. Molecular biologists know how to recognize genes as they scan through nucleotide sequences—they can recognize stop and start signals, and certain control regions associated with every gene. They can also tell a great deal about the nature of the protein a gene encodes by examining its gene sequence: whether the protein is likely to be embedded in a membrane (like a receptor, for example), or whether it is a good candidate for an enzyme, and even what general type of enzyme.

If a human gene sequence is particularly puzzling, it is almost always possible to fish the same gene out of a mouse, where its function can be more easily deduced. The gene could be knocked out, for example, or modified in some way to see how the host animal is affected. The ability to probe gene function in other species, as a means of gaining insight into human gene function, is one of the reasons that a significant portion of the Human Genome Project budget was set aside for sequencing the genomes of other animals, from yeast through mice.

We have come a very long way since Mendel first described the mysterious entities we now call genes a little over a hundred years ago. In fact, many of the genes he identified, such as those affecting pea shape and plant height, have been isolated, cloned, and sequenced, and the function of the corresponding gene products is known in great detail. Progress in the new molecular genetics over the next hundred years or so will be even more stunning. There is every reason to believe that during that time we will have isolated and characterized the principal genes underlying human behavior, and identified most of their important allelic variants. Sorting out how these genes interact with each other and with the environment to produce behavioral phenotypes will almost certainly be a major theme in human biology in the twenty-first century.

Appendix II

A Brief History of Eugenics

The strong feelings evoked by research in human genetics are attributable in part to the fact that from the very beginning, and for a great many years thereafter, there were two quite different faces to this undertaking. One was simply an inquiry into the nature and extent of the obvious differences in various human phenotypes, and the assessment of the contributions of genotypic variation and environmental variation to these differences. An important part of this effort was the development of increasingly sophisticated quantitative and statistical methods for interpreting existing data, and for sorting out genetic and environmental influences.

The other aspect of this research, which lies at the origin of most of the controversy surrounding human genetics, was a strong desire on the part of the founders of the field, and of a good many of their successors, to apply the results of this research to "the betterment of the human race," through either voluntary or socially enforced selective breeding.* These individuals were interested in eugenics as defined by Darwin's cousin, Francis Galton, which has enormous social and cultural implications. The inquiry into genetic versus environmental causes of human variation is still vigorously pursued in many respected universities and research institutions throughout the world. Without exception, Galtonian eugenics has been abandoned as a serious field of study.

---

* For a detailed treatment of the history of eugenics in Europe and the United States, the reader is referred to the excellent treatise by Daniel J. Kevles, *In the Name of Eugenics: Genetics and the Uses of Human Heredity*, 2nd ed. (Cambridge, Mass.: Harvard University Press, 1995).

Eugenics arose out of the new field of genetics at a time of great intellectual ferment in European and American science. The Enlightenment promise of an important role for secular knowledge in the betterment of humankind seemed well on its way to fulfillment in the latter half of the nineteenth century. Developments in biology and medicine, in particular, captured the popular imagination. The demonstration that many common diseases are caused by living microorganisms, rather than being divine retribution for human misbehavior, was a change in thinking that can scarcely be imagined. Together with developments in the relatively new fields of immunology and public health, the new view of infectious disease would have almost immediate effects on human health and longevity.

Although with less obvious immediate practical benefit, the revolution in thinking caused by Darwin's ideas on evolution challenged a great many commonly held beliefs about human origins, and about the precise role of divine guidance in human affairs. But "Darwin's Dangerous Idea"* would also change the face of biology forever, particularly when later combined with the rediscovery of Mendel's work on heredity. Suddenly there was a rational framework for interpreting events in the natural biological world whose understanding had previously relied on the not always infallible fonts of common sense and divine wisdom. These new ways of thinking would allow humans to take charge of their world and shape it in ways never before imaginable. It is perhaps not surprising, then, that some of them made wrong-headed starts in trying to reshape humanity itself.

It is easy, at a remove of a hundred years or so in time, to ridicule the very idea that one could or even should think about altering the natural composition of the human species by altering the breeding habits of selected individual human beings, but that is precisely what eugenics set out to do. We now know enough about biology in general, and genetics in particular, to know that, from a scientific point of view, this would be a hopeless venture. The notion that some individuals would ever voluntarily reduce their breeding efforts in favor of other members of the species runs completely contrary to the past several billion years of evolutionary experience. Most biologists in the last century, like nearly all Western people during that period, probably believed

---

* Daniel C. Dennett, *Darwin's Dangerous Idea* (New York: Simon & Schuster, 1995).

that human beings could be brought beyond such base behavior by proper reflection, that humans have cultural endowments that can be used to overcome base biological imperatives. There is absolutely no evidence that this is the case. Individuals may choose for a variety of reasons to curtail their own reproductive activities, but these reasons do not extend to notions of improving the human species through voluntary withdrawal of their own genes from the general pool.

Humans have also changed considerably, in the past seventy-five years or so, their view of their relationship with nature and with one another. What was little appreciated at the end of the nineteenth century was the way human beings behave as social animals; sociology as a formal discipline was still in its infancy.* Like many other animals, including our closest relatives, the other higher primates, we organize ourselves within the species into social groups based on commonly accepted principles of self-identification. Like our primate cousins, individual human groups strongly resist efforts to challenge or alter those principles, just as they strongly resist the intrusion of other humans, with other organizing principles, into an established group. Differentness is ferociously discouraged in primate societies; exclusion of "others" is a major organizing principle. Unfortunately, we humans tend to interpret our own particular organizing principles as some sort of universal measure of good, and formalize our beliefs in them through various social institutions. We will fight, if necessary, to keep those institutions stable, to prevent change.

The problem, of course, is that humans are highly variable not only within the species, but even within identifiable but dynamically changing subgroups or "cultures." It seems apparent that during the evolution of human beings, there was for a time sufficient geographical and reproductive isolation to allow emergence of relatively distinct subgroups. But once these subgroups established contact with one another, the concept of subgroups—sometimes referred to as "races"—became ambiguous, and has for several thousand years been essentially meaningless. Today humans move quite readily between and among subgroups—a major difference between us and other primates—and so not all members of a given subgroup will necessarily have precisely

* The first university department of sociology was created at the University of Chicago in 1892.

the same organizing principles, and certainly not the same genotypes. And therein lies the rub. But this is a considerably more sophisticated view of ourselves than was current in the latter half of the nineteenth century. And it was in the context of this state of relative ignorance of how human beings function in social groups, in particular their tendency to behave, often brutally, in an exclusionary manner, that eugenics flourished well into the present century.

Nearly all of the early founders of the field of human genetics, beginning with Francis Galton himself, strongly believed that the human species not only could, but should, be subject to improvement by selective breeding practices. It was clear at the time that animals could be selectively bred to enforce certain behavioral patterns, and there was little reason to think that humans would in principle be any different. What was less clear was exactly which traits ought to be selected for among humans, who should do the selecting, and how people were to be convinced to go along with such a grand scheme. There was a vague notion that improved "mental function" was a worthy goal to aim for, although no one had a very precise idea of what the elements of superior mental function might be. They nevertheless were convinced that variations in mental function per se were heritable. Galton himself certainly had no interest in perpetuating the inbreeding system of the British aristocracy of his day; he considered many of them incompetent buffoons. He was more concerned with establishing a meritocracy, promoting the breeding of individuals with demonstrated accomplishments in any of a number of fields. Of course, it turned out that most of the individuals he thought desirable for breeding stock came mostly from his own his social class.

Galton was an unabashed eugenicist until the very end, leaving most of his personal fortune to support continued research into human heredity through establishment and endowment of the Galton Laboratory for National Eugenics. Those who took up the mantle of Galton's initial efforts, principally Karl Pearson in England and Charles Davenport in the United States, continued along the same lines established by Galton, namely, amassing data concerning variabilities in a wide range of human traits, along with further refining statistical means for analyzing these data. The notion of making any aspect of biology quantitative was revolutionary at the time, and like any revolution it was resolutely resisted by the scientific establishment. Neither biologists or mathematicians saw much value in it. The power of statistics in biology and medicine is now recognized, but it was not until the 1970s, for example, that clinical trials were subjected to the sort of rigorous statistical analysis necessary to guarantee the data will be meaningful. "Biostatisticians" are now routinely consulted *before* a clinical trial begins, to make sure the results of the trial will lead to useful conclusions. Galton and Pearson spearheaded this trend with the establishment, in 1902, of the professional journal *Biometrika*.

Like Galton, Pearson believed strongly in the possibility of improving the race through selective breeding. Of course, what he, too, largely had in mind was selective breeding of people in his own intellectual and social subgroup. It was Pearson who convinced Galton that reversion to the mean would not be an obstacle to improvement of the race, and that the mean itself could indeed be moved "forward" over time. Not surprisingly, Pearson became the first Galton Eugenics Professor in the Galton National Laboratory. In addition to his scientific studies, Pearson was much involved in some of the social experimentation of the time, particularly in regard to liberalization of Victorian society's attitudes toward women and sex. Darwinism not only challenged doctrinaire religious beliefs about the origin of human beings, but also stimulated discussions about their social evolution as well, with concepts such as "survival of the fittest" in a cultural as well as biological sense merging into what became known as "social Darwinism." These ideas were widely written about and discussed in intellectual circles around the turn of the century.

In the United States, Charles Davenport received a private endowment to establish the Eugenics Records Office as an adjunct to the Cold Spring Harbor Laboratory in New York. This laboratory would serve both as a repository for huge numbers of heredity questionnaires being gathered from various sources around the country and as a genetics research laboratory. To a large extent it was inspired by and became the American equivalent of the Galton Laboratory for National Genetics in England. Davenport had been among the first to demonstrate Mendelian (essentially, single gene) inheritance of a human trait, namely eye color.* A major debate at the time revolved

* Although eye color in humans can be adequately described on the basis of a single gene with two alleles (blue vs. brown), the wide range of color variants among nonblue eyes suggests the existence of a number of dark-eye "modifier" genes at work as well.

around the extent to which human traits in general would be inherited as variants of single genes, and which would be genetically more complex. Despite a few early successes, and the demonstration that certain diseases such as Huntington's chorea and hemophilia are controlled by single genes, it rapidly became apparent that most of the phenotypes of interest in humans would be genetically very complex—what we would now refer to as quantitative traits. But it must be admitted that in the early days of genetics this was simply not clear.

Like Pearson and Galton, Davenport was a determined eugenicist, and he tried hard to find a basis in the data he was collecting, and in his not inconsiderable knowledge of genetics, to promote eugenic thinking in the United States. His belief in a Mendelian basis of human inheritance led him to think that many undesirable traits, such as alcoholism or insanity, could fairly readily be "bred out" of the human population with a little eugenic guidance. Others, under the leadership of Thomas Hunt Morgan, had investigated simpler organisms such as Drosophila and found large numbers of traits controlled by single genes, or at least affected by single genes. It was not clear in the early days of human genetics to what extent heritable human characteristics might also be Mendelian in nature. No one pushed this point harder than Davenport, and his failure to provide convincing evidence for Mendelian inheritance of the more complex human behaviors set the tone for human genetics by subsequent workers.

The work of Galton and his immediate successors, and the increasing focus on heritability of mental traits, created a demand for better mental aptitude tests, and increasingly sophisticated statistical ways to interpret them. This need ultimately gave rise to a new branch of psychology called measurement psychology, or psychometrics. New tests, directed much more toward cognitive aspects of mental ability, arose through the work of Alfred Binet and his colleagues in France early in the twentieth century. Binet was interested in predicting the success of children in school based on individual tests, and he introduced concepts of verbal ability, cultural awareness, and spatial pattern recognition.

Implicit already in Binet's work was the notion that what he and others referred to as "intelligence" is likely to be composed of numerous subskills. Nevertheless, results from many different studies showed that in tests specifically structured to probe multiple aspects of mental function, individuals who scored high on one subcomponent of such a test tended to score well on other portions of the test as well. This led Charles Spearman, a British psychologist, to propose the existence of a single, heritable quality controlling overall mental ability, which he called the "g factor." The debate over whether there is a single, overarching factor involved in human mental function, how it is best measured, and whether or how such a factor is inherited, continues to this day. It is also important to realize that such tests were never designed to predict why certain people become violinists and others become physicists, and whether this is heritable, but only the overall level and heritability of mental function. Yet from a biological point of view, thinking of humans as they must have lived in the wild in prehistoric times, it may well be that the former question is more important than the latter.

Mental testing remained largely a research tool in most countries, but in the United States, like so many other things, it came to be mass-produced. By the 1920s, mental aptitude tests were applied across enormous segments of the American population, largely to school children and the military. Nearly two million soldiers were tested during the First World War. Tests by this time included mathematical as well as verbal reasoning. Children were shunted into various educational "tracks," depending on assumptions about their ability based on standardized tests. Various forms of testing for mental ability (along with personality assessment tests) eventually moved into the workplace as well, particularly in the United States.

Continued attempts to quantitate mental ability resulted in the gradual development and refinement of what have come to be known as "IQ tests." IQ is an acronym for "intelligence quotient," and as originally proposed by Wilhelm Stern in 1916, it was to be derived by dividing the mental age of a child (as determined by standardized tests) by his or her calendar age. In adults, this definition is inadequate; a fifty-year-old individual with an IQ of 150 may not enjoy being compared mentally with a seventy-five-year-old. Numerous other "mental aptitude" tests became available to provide the equivalent of IQ scores, one of the most common being the Wechsler Adult Intelligence Scale. Such scales look at where an individual ends up on a standard bell-shaped curve representing the performance of the population as a whole. Some psychometricians would like to believe that mental aptitude tests as presently constituted and administered can, if properly interpreted, provide a reasonable approximation of Spearman's famous g factor, but this point of view is by no means universal among psychologists.

By the 1930s, serious concern began to be expressed about the insistent focus on genetics and heritability in interpreting the results of mental testing. Claims that the quality of home environment and access to education could affect performance on standardized tests were increasingly recognized as valid. Particularly in the military, where many recruits were barely literate, performance on a linguistically complex test (both its contents and its accompanying instructions) could hardly be a measure of native intelligence. Yet the eugenics movement continued to grow, both in the United States and Europe, aided by lay groups such as the American Eugenics Society, established in the United States in 1926.

Eugenicists—both lay and scientific—sorted themselves out in a number of ways, reflecting their position in the societies in which they lived, and their own political, social, and religious beliefs. Eugenics was embraced by both liberals and conservatives of their day—it was, at least initially, not a political issue. Both atheists and religious leaders could be found among its supporters. What tied eugenicists together originally was that they were mostly intellectuals, interested more in ideas than in actions. The common concern among all eugenicists was the perceived under-reproduction of the fit—what we would call the "high-achievers"—and over-reproduction of the "unfit." A commonly heard phrase throughout the entire eugenics period was that the "bottom" one-quarter of each generation was producing three-quarters of the offspring of the succeeding generation; the disastrous consequences of continuing this practice were judged to be self-evident.

Eugenicists tended to be either positive or negative in their approaches to correcting the perceived degradation of the "race" through unregulated reproduction. Positive eugenicists felt that the best approach to dealing with the problem of under-reproduction of the able and the fit was to try to convince this group to be more reproductively active. Already in the nineteenth century the trend, independent of eugenic guidance, was for more educated, more economically successful people to have fewer children. Positive eugenicists thought such people should be educated to their "duty to their race" to produce more children, and if that didn't work they should be enticed with various economic incentives, through tax breaks or outright financial rewards. Negative eugenicists seemed more focused on the undesirable properties of those they regarded as unfit, and looked for ways to limit their numbers, using whatever means and powers necessary to achieve that end. The "unfit" rapidly became a catch-all category for anyone significantly different from the individuals doing the categorizing. Although originally intended to describe underachievers in general, and in particular the truly mentally handicapped, it soon took on class, racial, and ethnic tones. The categorizers were almost always from the educated white classes, usually formally educated but occasionally self-educated or "self-made"; the unfit were almost always uneducated, unskilled, or persons of color.

In the United States, with its rapidly expanding influx of different immigrant types, one of the means preferred by negative eugenicists was to limit the numbers of the unfit through more rigorous control of immigration. Certain ethnic groups—especially those from Asia and Southern and Eastern Europe—were suspected of having innately high levels of mental incompetence, criminal tendencies, and other undesirable traits. Pro-eugenic scientists, particularly those in Charles Davenport's Eugenics Records Office, culled their data bases to provide government officials with the necessary supportive evidence for these assertions. IQ tests were used to show that the common wisdom about immigrants was true. Barely literate in English, even after several years in America, they of course scored in the lower percentile ranges.

That immigrants in fact actually possessed the low-brow traits they were suspected of, at least to a higher degree than the generally preferred Northern European stock, was entirely unclear. Even less clear at the time was which of these traits, assuming someone actually possessed them, were heritable. Rather than recognizing the extraordinary courage and determination involved in uprooting self and family from one culture and moving to another, and the possible value of the underlying heritable traits this might represent, xenophobic obsessions with the perceived unfitness of foreign immigrants found expression in increasingly harsh and exclusionary immigration laws. In the United States such laws began to be formulated in the late nineteenth century. Initially intended mostly for Asians, by the 1920s exclusionary acts were aimed basically at anyone who was not white, Anglo-Saxon, and Protestant. The exclusionary tendency reached its zenith in the United States with the infamous Immigration Act of 1924.

Another form of negative eugenics involved direct interference in the reproductive activities of those perceived to be unfit. Various plans were put forth to promote birth control among the undesirable classes, but opposition from religious groups, the costs involved, and problems of compliance made these schemes unworkable. Persons with known or suspected heritable diseases were strongly counseled not to marry at all, or at least not to marry each other, and in many states laws were passed to make genetically undesirable marriages illegal. This could be viewed as relatively benign, but the evidence that some of these diseases were truly heritable was actually quite weak at the time, and certainly could not have provided a scientifically valid basis for state interference in individual marriage contracts. For the severely mentally handicapped, the "criminally insane," and certain types of repeat criminals, laws were sought to compel sterilization, particularly of institutionalized individuals. Unlike immigration laws, which were the province of the federal government in the United States, laws regulating marriage and procedures such as sterilization are left to individual states. At one point more than thirty states had laws of this kind on their books, and some still do today. The legality of such laws was even upheld by the U.S. Supreme Court, which compared a state's right to compel sterilization to compulsory vaccination!

Although state interference in such matters as marriage and sterilization were more widespread in some of the Scandinavian countries* than in the United States or Britain, eugenics as an official state policy was carried to its greatest extremes in Nazi Germany. Prior to the ascendancy of the National Socialist government in 1933, Germany was not particularly actively involved in eugenics, either as an official state concern or as a major organizing principle of private organizations. German scientists had naturally followed the development of eugenics in Britain and the United States with great interest, but had made few significant contributions of their own. But as soon as the Nazis came to power, a number of both positive and negative eugenics policies were put into effect almost immediately. Families of certi-

---

* In 1999, the Swedish government approved financial compensation for survivors among the estimated 63,000 people—mostly women—sterilized between 1936 and 1976 for reasons of perceived "genetic inferiority."

fied Aryan stock were provided with a range of incentives for producing as many children as possible. Negative eugenics was practiced on a scale well beyond anything that had taken place in other countries. A law not only legalizing but making compulsory sterilization of the unfit was passed in 1933—the so-called "Law for the Prevention of Genetically Diseased Progeny." Doctors were required to report persons with mental or physical handicaps that were deemed even remotely heritable. Since the genetic basis of most of the defects involved was poorly understood at the time, there was not even the slightest scientific justification for the actions carried out. But these were politically, not scientifically, motivated programs. Over 300,000 people were sterilized by the time Germany initiated the Second World War. More very likely would have been, but by that time those considered defective, including eventually anyone of Jewish descent, were simply killed in the extermination camps.

The history of eugenics in Europe and the United States contains within it many sad lessons, but perhaps the most important of these for science is what can happen when social and political motives or personal ideologies drive scientific inquiry, rather than the desire for clear answers to fundamental scientific questions. The twentieth century has seen other examples of science derailed for social and political reasons; the disastrous influence of Lysenko on Soviet genetics is but one example. But just as we must not yield to the temptation to use science to achieve nonscientific ends, we must also not let revulsion against abuses carried out in the name of science divert us from the quest to uncover underlying truths, as best we can understand them. Many very credible scientists to this day oppose a good deal of what has been learned about human genetics, not so much because the data are wrong, but because they fear—not without some justification, given the history of the field—the implications of these data for humanity. Their fears may or may not be justified, but that is a question that must be addressed on political, not scientific, grounds.

Certainly the horrors of the Nazi eugenic programs, once they could no longer be denied in England and America, put an immediate end to ideas about applying knowledge of human genetics to the solution of social and political problems. Unfortunately, it also greatly dampened enthusiasm for research in human genetics generally. Opposition to eugenics had been building for some time—not surprisingly, among the groups targeted as unfit by those influencing eugenic thinking. In the United States, this tended to be mostly people who were not white, Anglo-Saxon Protestants. But the excluded and the unfit formed some notable constituencies, such as the Catholic Church, Jewish groups, and private as well as public welfare agencies that acted as advocates for the poor and the disadvantaged—another major component of society regarded by some as de facto "unfit." Resistance from women's rights groups, made up of many of the types of women positive eugenicists were encouraging to bear increased numbers of children, also continued to mount.

All of these attacks and more, based on the social and political aspects of eugenics, did not challenge the underlying science. But a number of scientists began to challenge eugenics as well. In any field of science there is always vigorous, and sometimes acrimonious, debate over almost any issue worth addressing. A good many scientists, of course, spend most of their time trying to find evidence to support mainstream theories, because they want to be part of the mainstream. That was certainly true of eugenics for a number of years. But no scientific paradigm remains intact for long; scientific theories are ever-shifting, ever-evolving entities, and they evolve by constant experimentation and subsequent debate among the experimenters. From Galton's time forward, the data on human heredity was under the constant scrutiny and attack that characterizes any scientific inquiry, and there were always other interpretations of the data.

The downfall of eugenics in scientific circles did not emerge from the generation of independent data disproving a role of genetics in the variability of human behavior. The contention of Davenport and his supporters that most human mental and personality traits would be inherited in a Mendelian fashion was rather quickly knocked down, but the basic notion that variability in these traits was to some degree heritable—a major contention of the eugenicists—was not. The notion that many heritable human traits are quantitative, rather than Mendelian, in nature was well understood by 1930. Most scientists who attacked eugenics did so because they realized that the data on human heredity in general were far from being sufficiently strong to serve as a basis for guiding *any* social or political programs, and because they could see quite clearly that eugenics was rapidly becoming the driving force for racially and ethnically oriented programs of social exclusion and oppression. They may well have disagreed with some of the underlying science, but that is not what drove them to oppose it so vigorously.

A New Eugenics?

Over the past twenty years, there has been an enormous increase in our understanding of the genetic basis of many diseases, and in the technology for isolating and manipulating the particular alleles of the genes responsible for these diseases. This has made it possible to determine, for example, whether someone at risk for a genetic disease (based on family history) actually carries the gene. In the case of diseases like cystic fibrosis, which is caused by a recessive gene, it is possible to be a carrier of the disease without experiencing any of the symptoms. But when two carriers begin producing children, on average one in four offspring will inherit two defective alleles of the underlying gene, and develop cystic fibrosis (which is caused by a defect in a cell-surface ion channel.)

But our evolving understanding of genetic diseases, together with recent advances in reproductive technology, has made it possible to do more than just detect carriers of diseases like cystic fibrosis. It is now possible, at least in theory, to eliminate such genes forever from the human genome. The technique of in vitro fertilization was developed to assist couples who have had difficulty in establishing conception through normal intercourse. In vitro fertilization involves collection of sperm and ova from the prospective parents, and mixing them together in the laboratory under conditions which favor fertilization—penetration of an egg by a sperm. In most cases, multiple fertilizations occur during this in vitro process, generating numerous *zygotes* (the cells formed as a result of fertilization). The zygotes are normally monitored for several days, during which they undergo the initial rounds of cell division that will ultimately produce a complete human being. Zygotes that successfully begin the process of cell division are referred to as *embryos*. Those embryos that begin this division process normally, and appear normal under the microscope, are potentially available for implantation into the prospective mother.*

---

* For more details on recent advances in reproductive technology, see R. Gosden, *Designing Babies* (New York: W. H. Freeman, 1999).

In those cases where one or both of the prospective parents are known to carry a potentially harmful gene, it is now possible through standard genetic techniques to detect the presence of this gene either in the mother's eggs, prior to fertilization in vitro, or in the cells of the early embryo. The physician performing the in vitro fertilization, in consultation with the parents, can then decide which eggs to fertilize and reimplant in the mother, or, after fertilization, determine which of the embryos is carrying two bad alleles of the gene. The unused eggs—or defective embryos—are then simply discarded.

It is obvious why this approach (which has already been used numerous times to produce disease-free infants) raises ethical concerns. While it may seem difficult to argue with sparing an unformed child the suffering that can accompany an inherited disorder like cystic fibrosis, the techniques involved could easily be turned to other uses. In just a few more years, as a result of the Human Genome Project, we will have a complete catalog of every gene in the human genome, as well as all of the variants of these genes found in the population. As the functions of these genes are sorted out, and the alterations associated with their various alleles are understood, will the "techno-couple" of the future and their "gene doctor" begin to select eggs or embryos on other bases, such as physical appearance or personality traits? And what is to stop us from *adding* genes to eggs or embryos? Will future parents demand that they be allowed to scan through a catalog of available human genes, shopping for alleles they would like to see in their children? Will the human equivalent of the mouse NMDA gene seem like a sure bet for improving IQ? This is by no means a fantasy; the technology already exists. Biologists have been producing strains of genetically tailored mice, through addition of foreign genes, for over a decade.

We should not pass too lightly over the fact that genetic manipulations carried out at the reproductive level do not affect just a single individual; they may reach far into the future as well. In the case of discarding embryos with known disease-causing alleles of certain genes, we are in effect purging that allele from all future members of the family involved. In the case of alleles that cause some catastrophic diseases, like Huntington's disease, it is difficult to make a case that this should not be done. But should we view diseases like cystic fibrosis in the same way? What are the consequences of purging a particular form of a gene from an entire population, if we could do that? It turns out that the frequency of some of the alleles of the gene causing cystic fibrosis are inexplicably high in certain populations. Why is that? We would expect alleles of a gene that affects reproductive ability (as the cystic fibrosis alleles do) to rapidly disappear from a population. Could there be some unperceived advantage of the presence of these alleles, at least heterozygotically, in some populations? In the case of the variants of the gene that causes sickle-cell anemia, we know that these alleles confer protection against malaria in heterozygotes (one normal and one defective globin gene), which certainly could account for their very high frequency in African peoples and their descendants. Was there in the past, or is there even now, some similar advantage to the alleles of the gene causing cystic fibrosis that we don't understand? Given that we can screen for this gene in at-risk individuals and provide them with reproductive counseling, and that we may well be able to correct the underlying defect by gene therapy in those with active disease, ought we to rush into chasing this allele out of the human species for all time?

The possibility that we may be able to purge harmful alleles of certain genes from the human species raises the specter of defining a new category of the "genetically unfit" in human society, and has brought some to raise the red flag of eugenics before us once again.* In the case of at least some genetic diseases, there would be little argument against ridding humanity of the underlying alleles that bring unacceptable misery and suffering to afflicted individuals and their families. But what if we eventually find genetic alleles that are responsible for poor performance on an IQ test? What if we define the alleles of other genes that play a major role in aggressiveness, or criminality, or homosexuality? Almost certainly there will be at least a few individuals who will want to use this new information to manage their own reproductive affairs. Past history tells us that if such people gain political power,

* Even the American Society for Human Genetics, a professional organization supporting and promoting human genetic research, was sufficiently concerned about these developments that it recently issued a cautionary statement in its official journal. See *American Journal of Human Genetics* 64:335–338 (1999).

they may also try to impose their views on societies as a whole. How do we stop that from happening? We should—we must—think very seriously about this, or we stand to lose all of the benefits that a better understanding of our genetic selves could bring.

References

Chapter 1

Bouchard, T. 1994. Genes, environment and personality. *Science* 264:1700–1701.

Bouchard, T., and P. Propping, eds. 1993. *Twins as tools of behavioral genetics.* Dahlem Workshop Report no. 53. New York: J. Wiley and Sons.

Bouchard, T., et al. 1990. Sources of human psychological differences: The Minnesota study of twins reared apart. *Science* 250: 223–228.

DiLalla, D., et al. 1996. Heritability of MMPI personality indicators of psychopathology in twins reared apart. *Journal of Abnormal Psychology* 105:491–499.

Hur, Y., and T. Bouchard. 1997. The genetic correlation between impulsivity and sensation-seeking traits. *Behavior Genetics* 27:455–463.

Lykken, D., et al. 1993. Heritability of interests: A twin study. *Journal of Applied Psychology* 78:649–661.

McClearn, G., et al. 1997. Substantial genetic influence on cognitive abilities in twins eighty or more years old. *Science* 276:1560–1563.

Waller, N., et al. 1990. Genetic and environmental influences on religious interests, attitudes and values. *Psychological Science* 1:138–142.

Wright, L. 1997. *Twins: What they tell us about who we are.* New York: J. Wiley & Sons.

Chapter 2

Haynes, W., et al. 1998. The cloning by complementation of the pawn-A gene in *Paramecium. Genetics* 149:947–957.

Hinrichsen, R., Y. Saimi, and C. Kung. 1984. Mutants with altered Ca^{++} channel properties in *Paramecium tetraurelia*: Isolation, characterization and genetic analysis. *Genetics* 108:545–558.

Jennings, H. 1906. *Behavior of lower animals*. Bloomington: Indiana University Press.

Kink, J., et al. 1990. Mutations in *Paramecium* calmodulin indicate functional differences between the C-terminal and N-terminal lobes in vivo. *Cell* 62:165–174.

Saimi, Y., et al. 1983. Mutant analysis shows that the Ca^{++}-induced K^{+} current shuts off one type of excitation in Paramecium. *Proceedings of the National Academy of Science* 80:5112–5116.

Chapter 3

Abbott, D., et al. 1993. Specific neurendocrine mechanisms not involving generalized stress mediate social regulation of female reproduction in cooperatively breeding marmoset monkeys. *Annals of the New York Academy of Sciences* 807:219–238.

Belluscio, L., et al. 1999. A map of pheromone receptor activation in the mammalian brain. *Cell* 97:209–220.

Halpern, M. 1987. The organization and function of the vomeronasal system. *Annual Review of Neuroscience* 10:325–362.

Herrada, G., and C. Dulac. 1997. A novel family of putative pheromone receptors in mammals with a topographically organized and sexually dimorphic distribution. *Cell* 90:763–773.

McClintock, M. 1971. Menstrual synchrony and suppression. *Nature* 229:244–245.

McClintock, M. 1984. Estrus cycle: Modulation of ovarian cycle length by female pheromones. *Physiology and Behavior* 32:701–705.

McClintock, M. 1998. On the nature of mammalian and human pheromones. *Annals of the New York Academy of Sciences* 855:390–392.

Monti-Bloch, L., et al. 1998. The human vomeronasal system. *Annals of the New York Academy of Sciences* 855:373–389.

Ober, C., et al. 1997. HLA and mate choice in humans. *American Journal of Human Genetics* 61:497–504.

Porter, R., and J. Winberg. 1999. Unique salience of maternal breast odors for newborn infants. *Neuroscience and Biobehavioral Reviews* 439–449.

Preti, G., et al. 1987. Human axillary extracts: Analysis of compounds from samples which influence menstrual timing. *Journal of Chemical Ecology* 13:717–722.

Schank, J., and M. McClintock. 1997. Ovulatory pheromone shortens ovarian cycles of rats living in olfactory isolation. *Physiology and Behavior* 62:899–904.

Sorensen, P., et al. 1998. Discrimination of pheromonal clues in fish: Emerging parallels with insects. *Current Opinions in Neurobiology* 8:458–467.

Stern, K., and M. McClintock. 1998. Regulation of ovulation by human pheromones. *Nature* 392:177–179.

Tirindelli, R., et al. 1998. Molecular aspects of pheromonal communication via the vomeronasal organ of mammals. *Trends in Neuroscience* 21:482–486.

Wedekind, C. 1997. Body odour preferences in men and women: do they aim for specific MHC combinations or just heterozygosity? *Proceedings of the Royal Society of London* Series B, 264:1471–1479.

Weller, L., and A. Weller. 1993. Human menstrual synchrony: a critical assessment. *Neuroscience and Biobehavioral Reviews* 17:427–439.

Chapter 4

Bailey, C., D. Bartsch, and E. Kandel. 1996. Toward a molecular definition of long-term memory storage. *Proceedings of the National Academy of Sciences* 93:13445–13452.

Bargmann, C. 1993. Genetic and cellular analysis of behavior in *C. elegans*. *Annual Review of Neuroscience* 16:47–71.

Bargmann, C. 1998. Neurobiology of the *C. ele*gans genome. *Science* 282:2028–2032.

C. elegans Sequencing Consortium. 1998. Genome sequence of the nematode *C. elegans*: A platform for investigating biology. *Science* 282:2012–2018.

De Bono, M., and C. Bargmann. 1998. Natural variation in a neuropeptide Y receptor homolog modifies social behavior and food responses in *C. elegans*. *Cell* 92:217–227.

Frost, W., et al. 1985. Monosynaptic connections made by the sensory neurons of the gill and siphon withdrawal reflex in *Aplysia* participate in the storage of longterm memory for sensitization. *Proceedings of the National Academy of Sciences* 82:866–869.

Kandel, E. 1979. *Behavioral biology of Aplysia*. New York: W. H. Freeman and Company.

Troemel, E., et al. 1997. Reprogramming chemotaxis responses: Sensory neurons define olfactory preferences in *C. elegans*. *Cell* 91:161–169.

Ware, R., et al. 1975. The nerve ring of the nematode *Caenorhabditis elegans*: Sensory input and motor output. *Journal of Comparative Neurology* 162:71–110.

Wen, J., et al. 1997. Mutations that prevent associative learning in *C. elegans*. *Behavioral Neuroscience* 111:354–368

Chapter 5

DeFries, J., et al. 1978. Response to 30 generations of selection for open-field activity in laboratory mice. *Behavior Genetics* 8:3–13.

Flint, J., et al. 1995. A simple genetic basis for a complex psychological trait in laboratory mice. *Science* 269:1432–1435.

Chapter 6

Campbell, S., and P. Murphy. 1998. Extraocular circadian phototransduction in humans. *Science* 279:396–399.

Crosthwaite, S., et al. 1997. Neurospora wc-1 and wc-2: Transcription, photoresponses, and the origin of circadian rhythmicity. *Science* 276:763–769.

Czeisler, C., et al. 1999. Stability, precision, and near-24-hour period of the human circadian pacemaker. *Science* 284:2177–2181.

Deeb, S., and A. Motulsky. 1996. Molecular genetics of color vision. *Behavior Genetics* 26:195–207.

Dunlap, J. 1996. Genetic and molecular analysis of circadian rhythms. *Annual Review of Genetics* 30:579–601.

Ishiura, M., et al. Expression of a gene cluster kaiABC as a circadian feedback process in cyanobacteria. *Science* 281:1519–1523.

King, D., et al. 1997. Positional cloning of the mouse circadian *clock* gene. *Cell* 89:641–653.

Klein, T., et al. Circadian sleep regulation in the absence of light perception: Chronic non-24-hour circadian rhythm sleep disorder in a blind man with a regular 24-hour sleep-wake schedule. *Sleep* 16:333–343.

Konopka, R., and S. Benzer. 1971. Clock mutants of *Drosophila melanogaster*. *Proceedings of the National Academy of Sciences* 68:2112–2116.

Miyamoto, Y., and A. Sancar. 1998. Vitamin B2-based blue-light photoreceptors in the retinohypothalamic tract as the photoactive pigments for setting the circadian clock in mammals. *Proceedings of the National Academy of Sciences* 95:6097–6102.

Ralph, M., et al. 1990. Transplanted suprachiasmatic nucleus determines circadian period. *Science* 247:975–978.

Rosato, E., et al. 1997. Circadian rhythms: From behavior to molecules. *BioEssays* 19:1075–1082

Tei, H., et al. 1997. Circadian oscillation of a mammalian homologue of the Drosophila *period* gene. *Nature* 389:512–516.

Chapter 7

Bailey, C., et al. 1996. Toward a molecular definition of long-term memory storage. *Proceedings National Academy of Science* 93:13445–13452.

Balling, A., et al. 1987. Are the structural changes in adult Drosophila mushroom bodies memory traces? Studies on biochemical learning mutants. *Journal of Neurogenetics* 4:65–73.

Bear, M. 1996. A synaptic basis for memory storage in the cerebral cortex. *Proceedings National Academy of Science* 93:13453–13459.

Cashmore, A., et al. 1999. Cryptochromes: Blue-light receptors for plants and animals. *Science* 284:760–765.

Dauwalder, B., and R. Davis. 1995. Conditional rescue of the dunce learning/memory and female fertility defects with Drosophila or rat transgenes. *Journal of Neuroscience* 15:3490–3499.

Davis, R. L., et al. 1995. The cyclic AMP system and *Drosophila* learning. *Molecular and Cellular Biochemistry* 149/150:271–278.

Davis, R., and B. Dauwalder. 1991. The Drosophila *dunce* locus. *Trends in Genetics* 7:224–229.

Dudai, Y., et al. 1976. Dunce, a mutant of Drosophila deficient in learning. *Proceedings of the National Academy of Sciences* 73:1684–1688.

Engert, F., and T. Bonhoeffer. 1999. Dendritic spine changes associated with hippocampal longterm synaptic plasticity. *Nature* 399:66–69.

Engels, P., et al. 1995. Brain distribution of four rat homologues of the Drosophila dunce cAMP phosphodiesterase. *Journal of Neuroscience Research* 41:169–78.

Feany, M., and W. Quinn. 1995. A neuropeptide gene defined by the Drosophila memory mutant amnesiac. *Science* 268:869–873.

Griffith, L., et al. 1993. Inhibition of calcium/calmodulin-dependent protein kinase in Drosophila disrupts behavioral plasticity. *Neuron* 10:501–509.

Hall, Jeffrey C. 1994. The mating of a fly. *Science* 264:1702–1714.

Imanishi, T., et al. 1997. Ameliorating effects of rolipram on experimentally induced impairments of learning and memory in rodents. *European Journal of Pharmacology* 321:273–278.

Lisman, J. 1994. The CaMII kinase hypothesis for the storage of synaptic memory. *Trends in Neurological Science* 17:406–412.

Liu, L., et al. 1999. Context generalization in *Drosophilia* visual learning requires the mushroom bodies. *Nature* 400:753–755.

Mayford, M., et al. 1996. Control of memory formation through regulated expression of a CaMKII transgene. *Science* 274:1678–1683.

Morimoto, B., and D. Koshland. 1991. Identification of cAMP as the response regulator for neurosecretory potentiation: A model memory system. *Proceedings of the National Academy of Sciences* 88:10835–10839.

Nassif, C., et al. 1998. Embryonic development of the Drosophila brain. I. Pattern of pioneer tracts. *Journal of Comparative Neurology* 402:10–31.

Neckameyer, W. 1998. Dopamine and mushroom bodies in Drosophilia: Experience-dependent and -independent aspects of sexual behavior. *Learning and Memory* 5:157–165.

Nighorn, A., et al. 1994. Progress in understanding the Drosophila *dnc* locus. *Comparative Biochemistry and Physiology* 108:1–9.

Phelps, E., and A. Anderson. 1997. Emotional memory: What does the amygdala do? *Current Biology* 7:R311–R314.

Siegel, R., et al. 1984. Genetic elements of courtship in *Drosophila*: Mosaics and learning mutants. *Behavior Genetics* 14:383–409.

Silva, A., et al. 1992. Impaired spatial learning in calcium-calmodulin kinase II mutant mice. *Science* 257:206–211.

Technau, G. 1984. Fiber number in the mushroom bodies of adult *Drosophila melanogaster* depends on age, sex and experience. *Journal of Neurogenetics* 1:113–126.

Weiner, J. 1999. *Time, love, memory* [Biography of Seymour Benzer]. New York: Alfred A. Knopf.

Yu, J., et al. 1997. Identification and characterization of a human calmodulin-stimulated phosphodiesterase PDE1B1. *Cellular Signaling* 9:519–529.

Chapter 8

Bellivier, F., et al. 1998. Serotonin transporter gene polymorphisms in patients with unipolar or bipolar depression. *Neuroscience Letters* 255:143–146.

Bliss, T. 1999. Young receptors make smart mice. *Nature* 401:25–26.

Boschert, U., et al. 1994. The mouse 5HT1b receptor is localized predominantly in axon terminals. *Neuroscience* 58:167–182.

Diagnostic and Statistical Manual of Mental Disorders. 1994. 4th ed. Washington, DC: American Psychiatric Association.

Galli, A., et al. 1997. Drosophila serotonin transporters have voltage-dependent uptake coupled to a serotonin-gated ion channel. *Journal of Neuroscience* 17:3401–3411.

Goldman, J., et al. 1996. Direct analysis of candidate genes in impulsive behaviors. *Ciba Foundation Symposium* 194:139–154.

Göthert, M., et al. 1998. Genetic variation in human 5HT receptors: Potential pathogenetic and pharmacological role. *Annals of the New York Academy of Science* 861:26–30.

Hollander, E., et al. 1998. Short-term single-blind fluvoxamine treatment of pathological gambling. *American Journal of Psychiatry* 155:1781–1783.

Jönsson, E., et al. 1998. Polymorphisms in the dopamine, serotonin, and norepinephrine transporter genes, and their relationships to monoamine metabolite concentrations in CSF of healthy volunteers. *Psychiatry Research* 79:1–9.

Kendler, S., and C. Prescott. 1999. A population-based twin study of lifetime major depression in men and women. *Archives of General Psychiatry* 56:39–44.

Lam, S., et al. 1996. A serotonin receptor gene (5HT1a) variant found in a Tourette's syndrome patient. *Biochemical and Biophysical Research Communications* 219:853–858.

Mason, S. 1984. *Catecholamines and behavior*. Cambridge: Cambridge University Press.

Nielsen, D., et al. 1992. Genetic mapping of the human tryptophan hydroxylase gene on chromosome 11 using an intronic conformational polymorphism. *American Journal of Human Genetics* 51:1366–1371.

Nierenberg, A., et al. 1998. Dopaminergic agents and stimulants as antidepressant augmentation strategies. *Journal of Clinical Psychiatry* 59 (Suppl. 5):60–63

Nöethen, M., et al. 1994. Identification of genetic variation in the human serotonin 1Db receptor gene. *Biochemical and Biophysical Research Communications* 205:1194–1200.

Olde, B., and W. McCombie. 1997. Molecular cloning and functional expression of a serotonin receptor from *Caenorhabditis elegans. Journal of Molecular Neuroscience* 8:53–62.

Quirarte, G., et al. 1998. Norepinephrine release in the amygdala in response to footshock. *Brain Research* 808:134–140.

Siever, L. 1997. *The new view of self: How genes and neurotransmitters shape your mind, your personality, and your mental health.* New York: Macmillan.

Tang, Y., et al. 1999. Genetic enhancement of learning and memory in mice. *Nature* 401:63–69.

Tsien, J., et al., 1996. The role of hippocampal CA1 NMDA receptor-dependent synaptic plasticity in spatial memory. *Cell* 87:1327–1338.

Chapter 9

Bonhomme, N., and E. Esposito. 1998. Involvement of serotonin and dopamine in the mechanism of action of novel antidepressant drugs. *Journal of Clinical Psychopharmacology* 18:447–452.

Boschert, U., et al. 1994. The mouse 5HT1b receptor is localized predominantly on axon terminals. *Neuroscience* 58:167–182.

Brunner, D., and R. Hen. 1997. Insights into the neurobiology of impulsive behavior from serotonin receptor knockout mice. *Annals of the New York Academy of Sciences* 836:81–105.

Brunner, H., et al. 1993. Abnormal behavior associated with a point mutation in the structural gene for monoamine oxidase A. *Science* 262:578–580.

Brunner, H., et al. 1993. X-linked borderline mental retardation with prominent behavioral disturbance: Phenotype, genetic localization and evidence for disturbed monoamine metabolism. *American Journal of Human Genetics* 52:1032–1039.

Coccaro, E., and R. Kavoussi. 1997. Fluoxetine and impulsive aggressive behavior in personality-disordered subjects. *Archives of General Psychiatry* 54:1081–1088.

Coccaro, E., et al. 1997. Heritability of aggression and irritability: A twin study of the Buss-Durkee aggression scales in adult male subjects. *Biological Psychiatry* 41:273–284.

Coccaro, E., et al. 1997. Serotonin function in human subjects: Intercorrelations among central 5-HT indices and aggressiveness. *Psychiatry Research* 73:1–14.

Dabbs, J., and M. Hargrove. 1997. Age, testosterone and behavior among female prison inmates. *Psychosomatic Medicine* 59:477–480.

DeVries, A., et al. 1997. Reduced aggressive behavior in mice with targeted disruption of the oxytocin gene. *Journal of Neuroendocrinology* 9:363–368.

Kavoussi, R., et al. 1997. The neurobiology of impulsive aggression. *Psychiatric Clinics of North America* 20:395–403.

Kriegsfield, L., et al. 1997. Aggressive behavior in male mice lacking the gene for neuronal nitric oxide synthase requires testosterone. *Brain Research* 769:66–70.

Lahn, B., and D. Page. 1997. Functional coherence of the human Y chromosome. *Science* 278: 675–683.

Marshall, J. 1998. Evolution of the mammalian Y chromosome and sex-determining genes. *Journal of Experimental Zoology* 281:472–481.

Maxson, S. 1996. Searching for candidate genes with effects on an agonistic behavior, offense, in mice. *Behavior Genetics* 26:471–476.

Nelson, R., and K. Young. 1998. Behavior in mice with targeted disruption of single genes. *Neuroscience and Biobehavioral Reviews* 22:453–462.

New, A., et al. 1998. Tryptophan hydroxylase genotype is associated with impulsive aggression measures. *American Journal of Medical Genetics* 81:13–17.

Ogawa, S., et al. 1997. Behavioral effects of estrogen receptor gene disruption in male mice. *Proceedings of the National Academy of Sciences* 94:1476–1481.

Renfrew, J. 1997. *Aggression and its causes: A biopsychosocial approach.* New York: Oxford University Press.

Seroczynski, A., et al. 1999. Etiology of the impulsivity/aggression relationship: Genes or environment? *Psychiatry Research* 86:41–57.

Simon, N., and R. Whalen. 1986. Hormonal regulation of aggression: Evidence for a relationship among genotype, receptor binding and behavioral sensitivity to androgen and estrogen. *Aggressive Behavior* 12:255–266.

Staner, L., et al. 1998. Association between novelty seeking and the dopamine D3 receptor gene in bipolar patients: A preliminary report. *American Journal of Medical Genetics (Neuropsychiatric Genetics)* 81:192–194.

Vernon, P., et al. 1999. Individual difference in multiple dimensions of aggression: A univariate and multivariate genetic analysis. *Twin Research* 2:16–21.

Vilain, E., and E. McCabe. 1998. Mammalian sex determination: From gonads to brain. *Molecular Genetics and Metabolism* 65:74–84.

Chapter 10

Bouchard, C., and A. Tremblay. 1997. Genetic influences on the response of body fat and fat distribution to positive and negative energy balance in human identical twins. *Journal of Nutrition* 127:943S–947S.

Bray, G., C. Bouchard, and W. James, eds. 1998. *Handbook of Obesity*. New York: Marcel Dekker, Inc.

Bray, M., et al. 1999. Linkage analysis of candidate obesity genes among the Mexican-American population of Starr County, Texas. *Genetic Epidemiology* 16:397–411.

Collier, D., et al. 1997. Association between 5-HT2a gene promoter polymorphism and anorexia nervosa. *Lancet* 350:412.

Comuzzie, A., et al. 1997. A major quantitative locus determining serum leptin levels and fat mass is located on chromosome 2. *Nature Genetics* 15:273–276.

Curzon, G., et al. 1997. Appetite suppression by commonly used drugs depends on 5HT receptors, but not on 5HT availability. *Trends in Pharmacological Sciences* 18:21–25.

Goran, M. 1997. Genetic influences on human energy expenditure and substrate utilization. *Behavior Genetics* 27:389–399.

Gorwood, P., et al. 1998. Genetics and anorexia nervosa: A review of candidate genes. *Psychiatric Genetics* 8:1–12

Hewitt, J. 1997. The genetics of obesity: What genetic studies have told us about the environment. *Behavior Genetics* 27:353–358

Horwitz, B., et al. 1998. Adiposity and serum leptin increase in fatty (fa/fa) BNZ neonates without decreased VMH serotonergic activity. *American Journal of Physiology* 274 (6 pt. 1):E1009–1016

Jeong, K., and S. Lee. 1999. High-level production of human leptin by fed-batch cultivation of recombinant Escherichia coli and its purification. *Applied and Environmental Microbiology* 65:3027–3032.

Leibowitz, S. 1990. The role of serotonin in eating disorders. *Drugs* 39 (Suppl. 3):33–48.

Lonnqvist F., et al. 1999. Leptin and its potential role in human obesity. *Journal of Internal Medicine* 245:643–652.

Lowe, M., and K. Eldredge. 1993. The role of impulsiveness in normal and disordered eating. In W. McCown, et al., eds., *The impulsive client: Theory, research and treatment*. Washington, D.C.: American Psychological Association. Pp. 151–184.

Montague, C., et al. 1997. Congenital leptin deficiency is associated with severe early-onset obesity in humans. *Nature* 387:903–907.

Nielsen, D. A., et al. 1998. A tryptophan hydroxylase gene marker for suicidality and alcoholism. *Archives of General Psychiatry* 55:593–602.

Norman, R., et al. 1998. Autosomal genomic scan for loci linked to obesity and energy metabolism in Pima Indians. *American Journal of Human Genetics* 62:659–668.

Reed, D., et al. 1997. Heritable variation in food preferences and their contribution to obesity. *Behavior Genetics* 27:373–385.

Samaras, K., et al. 1999. Genetic and environmental influences on total-body and central abdominal fat: The effect of physical activity in female twins. *Annals of Internal Medicine* 130:873–882.

Tartaglia, L., et al. 1995. Identification and expression cloning of a leptin receptor. *Cell* 83:1263–1271.

Tecott, L., et al. 1995. Eating disorder and epilepsy in mice lacking 5HT2c serotonin receptors. *Nature* 374:542–546.

Weltzin, E., et al. 1994. Serotonin and bulimia nervosa. *Nutrition Reviews* 52:399–406.

Wurtman, R., and J. Wurtman. 1995. Brain serotonin, carbohydrate craving, obesity and depression. *Obesity Research* 4 (Suppl. 3):477S–480S.

Zhang, Y., et al. 1994. Positional cloning of the mouse *obese* gene and its human homologue. *Nature* 372:425–431.

Chapter 11

Andretic, R., et al. 1999. Requirement of circadian genes for cocaine sensitization in Drosophilia. *Science* 285:1066–1088.

Bardo, M., et al. 1997. Effect of differential rearing environment on morphine-induced behaviors, opioid receptors and dopamine synthesis. *Neuropharmacology* 36:251–259.

Carlezon, W., et al. 1998. Regulation of cocaine reward by CREB. *Science* 282:2272–2275.

Carr, L., et al. 1998. A quantitative trait locus for alcohol consumption in selectively bred rat lines. *Alcoholism, Clinical and Experimental Research* 22:884–887.

Crabbe, J. 1998. Provisional mapping of quantitative trait loci for chronic ethanol withdrawal severity in BxD recombinant inbred mice. *Journal of Pharmacology and Experimental Therapeutics* 286: 263–271.

Crabbe, J., et al. 1996. Elevated alcohol consumption in null mutant mice lacking 5HT1b serotonin receptors. *Nature Genetics* 14:98–101.

Crabbe, J., et al. 1999. Identifying genes for alcohol and drug sensitivity: Recent progress and future directions. *Trends in Neuroscience* 22:173–179.

Enoch, M., and D. Goldman. 1999. Genetics of alcoholism and substance abuse. *Psychiatric Clinics of North America* 22:289–299.

Gianoukalis, C., et al. 1996. Implication of the endogenous opioid system in excessive ethanol consumption. *Alcohol* 13:19–23.

Heath, A., et al. 1997. Genetic and environmental contributions to alcohol dependence in a national twin sample: Consistency of findings in men and women. *Psychological Medicine* 27:1381–1396.

Herz, A. 1998. Opioid reward mechanisms: A key role in drug abuse? *Canadian Journal of Pharmacology* 76:252–258.

Hyman, S. 1996. Addiction to cocaine and amphetamine. *Neuron* 16:901–904.

Jayanthi, L., et al. 1998. The C. elegans gene T23G5.5 encodes an antidepressant- and cocaine-sensitive dopamine transporter. *Molecular Pharmacology* 54:601–609.

Johnson, E., et al. 1996. Indicators of genetic and environmental influence in alcohol-dependent individuals. *Alcoholism: Clinical and Experimental Research* 20:67–74.

Johnson, E., et al. 1996. Indicators of genetic and environmental influences in drug abusing individuals. *Drug and Alcohol Dependence* 41:17–23.

Kendler, K., et al. 1992. A population-based twin study of alcoholism in women. *Journal of the American Medical Association* 268:1877–1882.

Kim, M., et al. 1998. LUSH odorant-binding protein mediates chemosensory responses to alcohol in Drosophila melanogaster. *Genetics* 150:711–721.

Koob, G., et al. 1998. Neuroscience of addiction. *Neuron* 21:467–476.

Lappalainen, J., et al. 1998. Linkage of antisocial alcoholism to the serotonin 5HT1B receptor gene in two populations. *Archives of General Psychiatry* 55:989–994.

Lin, N., et al. 1996. The influence of familial and non-familial factors on the association between major depression and substance abuse/dependence in 1874 monozygotic twin pairs. *Drug and Alcohol Dependence* 43:49–55.

Maes, H., et al. 1999. Tobacco, alcohol and drug use in eight- to sixteen-year-old twins: The Virginia Twin Study of Adolescent Behavioral Development. *Journal of Studies on Alcohol* 60:293–305.

Markel, P., et al. 1997. Confirmation of quantitative trait loci for ethanol sensitivity in long-sleep and short-sleep mice. *Genome Research* 7:92–99.

Matthes, H., et al. 1996. Loss of morphine-induced analgesia, reward effect and withdrawal symptoms in mice lacking the m opioid receptor. *Nature* 383:819–823.

McGue, M., et al. 1996. Genotype-environment correlations and interactions in the etiology of substance abuse and related behaviors. *National Institute of Drug Abuse Reports* 159:49–73.

Merikangas, K., et al. 1985. Familial transmission of depression and alcoholism. *Archives of General Psychiatry* 42:367–372.

Merikangas, K., et al. 1998. Familial transmission of substance abuse disorders. *Archives of General Psychiatry* 55:973–979.

Miner, L., and R. Marley. 1995. Chromosomal mapping of the psychomotor stimulant effects of cocaine in BxD recombinant inbred mice. *Psychopharmacology* 122:209–214.

Nielsen, D., et al. 1998. A tryptophan hydroxylase gene marker for suicidality and alcoholism. *Archives of General Psychiatry* 55:593–602.

Nestler, E., and G. Aghajanian. 1997. Molecular and cellular basis of addiction. *Science* 278:58–63.

Phillips, T., et al. 1998. Localization of genes mediating acute and sensitized locomotor responses to cocaine in BxD/Ty recombinant inbred mice. *Journal of Neuroscience* 18:3023–3034.

Pierce, J., et al. 1998. A major influence of sex-specific loci on alcohol preference in C57BL/6 and DBA/2 mice. *Mammalian Genome* 9:942–948.

Reich, T., et al. 1998. Genome-wide search for genes affecting the risk for alcohol dependence. *American Journal of Medical Genetics* 81:207–215.

Rocha, B., et al. 1998. Increased vulnerability to cocaine in mice lacking the serotonin-1β receptor. *Nature* 393:175–178.

Rocha, B., et al. 1998. Differential responsiveness to cocaine in C57BL/6J and DBA/2J mice. *Psychopharmacology* 138:82–88

Sigvardsson, S., et al. 1996. Replication of the Stockholm adoption study of alcoholism. *Archives of General Psychiatry* 53:681–687

Sora, I., et al. 1998. Cocaine reward models: Conditioned place preference can be established in dopamine- and in serotonin-transporter knockout mice. *Proceedings of the National Academy of Science* 95:7699–7704.

Tsuang, M., et al. 1998. Co-occurrence of abuse of different drugs in men. *Archives of General Psychiatry* 55:967–972.

Uhl, G., et al. 1998. Dopaminergic genes and substance abuse. *Advances in Pharmacology* 42:1024–1032.

Chapter 12

Bouchard, T. 1998. Genetic and environmental influences on adult intelligence and special mental ability. *Human Biology* 70:257–279.

Daniels, J., et al. 1998. Molecular genetic studies of cognitive ability. *Human Biology* 70:281–296.

Plomin, R., et al. 1994. DNA markers associated with high versus low IQ: The IQ quantitative trait loci project. *Behavior Genetics* 24:107–118.

Plomin, R., et al. 1997. *Behavioral Genetics*. New York: W. H. Freeman and Co.

Chapter 13

Dittmann, R., et al. 1992. Sexual behavior in adolescent and adult females with congenital adrenal hyperplasia. *Psychoneuroendocrinology* 17:153–170.

Erhardt, A., and H. Meyer-Bahlburg. 1981. Effects of prenatal sex hormones on gender-related behavior. *Science* 211:1312–1318.

Gottschalk, A., et al. 1995. Evidence of chaotic mood variation in bipolar disorder. *Archives of General Psychiatry* 52:947–959.

Green, R. 1985. Gender identification in children and later sexual orientation: Followup of 76 males. *American Journal of Psychiatry* 142:339–341.

Hamer, D., and P. Copeland. 1995. *The science of desire: the search for the gay gene and the biology of behavior*. New York: Simon and Schuster (Touchstone).

Hamer, D., et al. 1993. A linkage between DNA markers on the X chromosome and male sexual orientation. *Science* 261:321–327.

Hu, S., et al. 1995. Linkage between sexual orientation and chromosome Xq28 in males but not females. *Nature Genetics* 11:248–256.

Li, L., and D. Hamer. 1995. Recombination and allelic association in the Xq/Yq homology region. *Human Molecular Genetics* 4:2013–2016.

Macke, J., et al. 1993. Sequence variation in the androgen receptor gene is not a common determinant of male sexual orientation. *American Journal of Human Genetics* 53:844–852.

Pattatucci, A. 1998. Molecular investigations into complex behavior: Lessons from sexual orientation studies. *Human Biology* 70:367–386.

Pillard, R., and J. Bailey. 1998. Human sexual orientation has a genetic component. *Human Biology* 70:347–365.

Rice, G., et al. 1999. Male homosexuality: Absence of linkage to microsatellite markers at Xq28. *Science* 284:665–669.

Risch, N., et al. 1993. Male sexual orientation and genetic evidence. *Science* 262:2063–2064 (see p. 2065 for Hamer's response).

Schüklenk, U., et al. 1997. The ethics of genetic research on sexual orientation. *Hastings Center Report* 27:6–13.

Sell, R., et al. 1995. The prevalence of homosexual behavior and attraction in the United States, the United Kingdom, and France. *Archives of Sexual Behavior* 242:35–248.

Vasey, P. 1995. Homosexual behavior in primates: A review of the evidence and theory. *International Journal of Primatology* 16:173–204.

Wallen, K., and W. Parsons. 1997. Sexual behavior in same-sex nonhuman primates: Is it relevant to human homosexuality? *Annual Review of Sex Research* 8:195–221.

Zucker, K., et al. 1996. Psychosexual development of women with congenital adrenal hyperplasia. *Hormones and Behavior* 30:300–318.

Chapter 14

Aitken, P., et al. 1995. Looking for chaos in brain slices. *Journal of Neuroscience Methods* 59:41–48.

Alper, J. 1995. Biological influences on criminal behavior: How good is the evidence? *British Medical Journal* 310:272–273.

Baxter, L., et al. 1992. Caudate glucose metabolic rate changes with both drug and behavior therapy for obsessive-compulsive disorder. *Archives of General Psychiatry* 49:681–689.

Clausing, P., et al. 1997. Differential effects of communal rearing and preweaning handling on open-field behavior and hot-plate latencies in mice. *Behavioural Brain Research* 82:179–184.

Dawkins, R. 1982. *The extended phenotype.* Oxford: Oxford University Press.

Duke, D. 1992. Measuring chaos in the brain: A tutorial review of nonlinear dynamical EEG analysis. *International Journal of Neuroscience* 67:31–80.

Gleick, J. 1987. *Chaos: Making a new science.* New York: Penguin Books.

Harlow, H., et al. 1965. Total social isolation in monkeys. *Proceedings of the National Academy of Science* 54:90–97.

Hinde, R., and Y. Spencer-Boothe. 1971. Effects of brief separation from mother on rhesus monkeys. *Science* 173:111–118.

Hinde, R., et al. 1978. Effects of various types of separation experience on rhesus monkeys 5 months later. *Journal of Child Psychology and Psychiatry* 19:199–211.

Insel, T., et al. 1999. Oxytocin, vasopressin and autism: Is there a connection? *Biological Psychiatry* 45:145–157.

Kolb, B., and I. Whishaw. 1998. Brain plasticity and behavior. *Annual Review of Psychology* 49:43–64.

Lewontin, R., et al. 1984. *Not in our genes: biology, ideology, and human nature.* New York: Pantheon.

Miklos, G., and G. Rubin. 1996. The role of the genome project in determining gene function: Insights from model organisms. *Cell* 86:521–529.

Pascual, R., et al. 1996. Effects of preweaning sensorimotor stimulation on behavioral and neuronal development in motor and visual cortex of the rat. *Biology of the Neonate* 69:399–404.

Rabinovich, M., and D. Abarbanel. 1998. The role of chaos in neural systems. *Neuroscience* 87:5–14.

Schwartz, J., et al. 1996. Systematic changes in cerebral glucose metabolic rate after successful behavior modification treatment of obsessive-compulsive disorder. *Archives of General Psychiatry* 53: 109–113.

Skinner, B. F. 1972. *Beyond freedom and dignity.* London: Cape Publishing.

Van Mier, H., et al. 1998. Changes in brain activity during motor learning measured with PET: Effects of hand of performance and practice. *Journal of Neurophysiology* 80:2177–2199.

Wang, Z., et al. 1998. Voles and vasopressin: A review of molecular, cellular and behavioral studies of pair bonding and paternal behaviors. *Progress in Brain Research* 119:483–498.

Wilson, E. O. 1975. *Sociobiology: The new synthesis.* Cambridge, Mass.: Harvard University Press.

Wilson, E. O. 1998. *Consilience.* New York: Alfred A. Knopf.

Young, L., et al. 1999. Increased affiliative response to vasopressin in mice expressing the V_{1a} receptor from a monogamous role. *Nature* 400:766–768.

Appendix I

Clark, W. 1997. *The new healers: The promise and problems of molecular medicine in the twenty-first century.* New York: Oxford University Press.

Plomin, R., et al. 1997. *Behavioral genetics.* New York: W. H. Freeman.

Appendix II

American Society for Human Genetics. 1999. Statement: Eugenics and the Misuse of Genetic Information to Restrict Reproductive Freedom. *American Journal of Human Genetics* 64:335–338.

Gosden, R. *Designing babies*. 1999. New York: W. H. Freeman.

Kevles, D. 1995. *In the name of eugenics*. Cambridge, Mass.: Harvard University Press.

Lewontin, R., et al. 1984. *Not in our genes: Biology, ideology, and human nature*. New York: Pantheon.

Index

Accessory olfactory bulb, 47
Acetylcholine, 67, 139
Addiction, 142, 199–202
  in rats, 206
Adoption studies, 149, 230, 232
Aggression, 142, 157–75, 187, 188, 213, 259
  breeding, 158–62
  female, 159, 163, 165, 167, 169, 172, 174
  impulsive vs. premeditated, 159, 164, 169
in mice, 163, 213
neurotransmitters, 167–75
pheromones and, 159
  predation, 158
  territory, 159
  testosterone in, 165–67
  verbal, 159
Aging, 77
Albinism, 88
Alcohol dehydrogenase, 212, 216
Alcoholism, 143, 146, 200, 201, 202, 210–18, 228, 284
  aggression and, 219
  Asian populations, 214
  drug addiction, relation to, 214
  environmental factors, 211
  gender bias, 211
  genetic components, 211
  psychotherapy for, 217
  type I vs. type II, 210, 211
  taste, role of, 213
  withdrawal, 213, 215
Alleles, 33, 81, 156, 169, 174, 175, 191, 192, 222, 236, 244–46, 249, 254, 258, 268, 271, 291
Allelic association studies, 237
Alzheimer's disease, 21, 75, 234–37
Afferent (sensory) neurons, 63
Alkaloids, 205
American Eugenics Society, 286
Amino acids, 139
γ-Amino butyric acid (GABA), 139
amnesiac mutation, 131, 255
Amphetamines, 147, 187, 189
Amygdala, 47, 131–32

Amyloid precursor protein, 234
Anabolic steroids, 165
Analgesia, 202
Anorexia nervosa, 143, 178, 188, 189
Ants, 43
apo gene, 235
app gene, 234
Aplysia californica, 118–124, 125, 127, 129–30, 134, 138, 150
Appetite control, 185
Associative conditioning, 50, 71
Astronauts (jet lag), 113
A-T, G-C pairing rule, 79
Attention-deficit disorder, 143, 200
Autism, 257, 262
Autonomic nervous system, 152
Autosomes, 78
Avoidance behavior, 30, 52
Axons, 66, 123, 141

Baboons, 53, 159
Bees, 43
Behavior
  avoidance, 30
  defined, 20
  reflexive, 34, 38, 64, 119–20, 204, 253
Barbiturates, 201
Bell curve, 89, 285
Benzer, Seymour, 102, 125, 126
Binet, Alfred, 284
Binge eating, 178
Biostatistics, 283
Bipolar disorder, 147
Blindness, 108, 113
Blood-brain barrier, 144, 184, 185
BMI. *See* Body-mass index
Body-mass index (BMI), 179, 190, 196
Bombykol, 43
Bouchard, Thomas, 13, 230
Boyse, Edward, 50
Brainstem, 206
Brenner, Sydney, 62, 125, 126
Bulimia nervosa, 143, 178, 188, 189

Cachexia, 178
Caenorhabditis elegans, 61–73, 93, 118, 119, 120, 123, 124–25, 253, 254, 256, 261, 266
Caffeine, 146
Calcium-calmodulin-dependent kinase, 134–35, 236
Calmodulin, 36, 38, 134–35
cAMP. *See* Cyclic AMP
cGMP. *See* Cyclic GMP
Cancer, 76
Carbohydrates, 180, 186, 194
Cardiovascular disease, 165
Cats, 43
ccg genes, 101
Cerebellum, 131–32, 145
Chaos, 268–70
Chemoreceptors, 42, 68
Cholesterol, 165, 235
Chromophores, 101, 108
Chromosome scan, 244
Chromosomes, 32, 89, 92, 124, 194, 271, 277
Cilia, 26, 30, 38, 46, 64

Circadian clocks, 96, 166, 210
Classical conditioning, 71, 123, 131
clock gene, 111, 210
Clocks, biological, 95–115, 150
  genes and, 111
  human, 96, 111
  location of, 105
  mammalian, 106–11
  metabolic regulation, 97
  phase difference, 96
  resetting, 100, 108, 110, 112
  temperature regulation, 97, 109
  temperature shifts, 106
Clones, 30
Cocaine, 146, 147, 201, 202, 207–10
  and alcohol abuse, 208
  tolerance, 208, 209
  withdrawal, 208
Codeine, 202
Cold Spring Harbor Laboratory, 283
Collaborative Study of Genetics of Alcoholism, 216
Color vision, 107
Colorado Adoption Study, 232
Cone cells, 107
Congenital adrenal hyperplasia, 247
Conidia (neurospora), 97
Conjugation, 31
Copy errors, 81, 83
Correlation, 12
Cortical arousal, 144
Cough center (brain), 203
Crick, Francis, 81, 125
Crossing over, 275–77
cry genes, 109
Cryptochrome, 108
Culture, 260, 281
Cutler, Winifred, 55
Cyclic AMP (cAMP), 122, 128, 129–30, 131, 151, 201, 207, 208, 209, 255
Cyclic GMP (cGMP), 256
Cystic fibrosis, 249, 291, 293
Cytokines, 42

Darwin, Charles, 28, 124, 225, 279
Darwinism, social, 283
Davenport, Charles, 282, 283, 290
db gene, 182
Delbrück, Max, 125n, 126
Delirium tremens, 214
δ opioid receptor, 205, 212
Dendrites, 66, 123, 141, 268
Dennet, Daniel, 280
Depression, 147–51, 155, 174, 187, 200, 259
Determinism, 265
Diabetes, 192, 193
Diet, 196
Diet-induced obesity, 180
Dihydroxytestosterone (DHT), 166
Dionne quintuplets, 8
DNA, 62, 77, 111, 193, 258, 270, 272
  and genes, 78
  double helix structure, 78
  markers, 10, 92, 194,

DNA (continued)
244, 272
nonsense, 93, 261, 270
segregation of, 28
sequencing, 277
Dopamine, 139, 149, 171, 189, 202, 204, 205, 206, 208, 209, 212
Dopamine transporter, 209
Drosophila melanogaster, 24, 111, 135, 253, 255, 261, 284
alcohol addiction in, 214
cocaine addiction in, 210
learning in, 124–30
mating behavior, 128–30
dunce mutation, 127–28, 129–30, 132, 255, 256
Dynorphin, 204–5.5, 207, 209

Eating disorders, 177–97
Eclosion, 102
Elephantiasis, 61
Embryo, 7, 277, 293, 294
β-Endorphin, 139, 205, 212
Enkephalins, 205
Enlightenment, the, 289
Environment, 6, 18, 90, 136, 137, 175, 221, 225, 262–65
in alcoholism, 211, 218–20
eating habits and, 179, 189
homosexuality, 241, 242
manipulation of, 17
nonshared, 12, 17
shared, 12, 17
Epinephrine, 139
Estradiol, 165
Estrogen, 165, 166
Estrus cycle, 48
Ethology, 24
Eugenics, 227, 230, 279–94
Eugenics Records Office, 283, 287
Euphoria, 146, 202, 203, 205, 206, 209
Exercise, 181, 190, 192, 196

Facilitating interneurons, 121, 123
Family studies (heritability), 11, 225, 242, 258
fast mutation, 34, **253**
Fat, 180, 181, 190, 192, 194, 196
Fearfulness, 86, 131, 51
Feeding behavior, 69
Fenfluramine, **189**
Fentanyl, 203
Fenfluramine, 187
Fenphen, 187
Fertilization, 291
Fluoxetine. *See* Prozac
forager mutation, 256
fmr gene, 236
Food preferences, 180
Fragile X syndrome, 236
Free will, 265
frq gene, 98

"g" factor, 285, 286
Galton, Francis, 28, 29, 225, 279, 282
Galton Laboratories for National Eugenics, 282
Gambling, 143, 200
Ganglia, 65, 67, 117, 125
Gaussian distribution, 226

Gender-atypical behavior, 247
Genes), 32–34, 62, 83, 92–94, 184, 256, 261, 264, 271–78. *See also individual gene names*
Genetic polymorphisms (alleles), 257, 273
Genetic privacy, 251
Genome, 32, 261, 272
Genome map, 92, 273
Genome scans, 94, 256, 257, 258
 alcoholism, 215–18
 mental ability, 237
 obesity, 193–95, 197
Genotype, 33, 90, 155, 191, 192, 218, 219, 222, 264, 275
Germ cells, 28, 60
Glucocorticoids, 53
Glutamate, 120, 139, 152
Glycine, 139

H-2, 50–52, 120
Habituation, 65, 121, 130
Hallucinogens, 200
Hamer, Dean, 239–52
Harris, Judith R., 232
Hemophilia, 242, 284
Heroin, 146, 202, 203, 206
Heterozygote advantage, 250, 293
Heterozygous, 84
Hippocampus, 131, 133–35, 152, 153
Histamine, 139
Histocompatibility, 50
HLA, 50
Homosexuality, 239–52, 293
 adopted siblings, 241
 environment, 241, 242
 evolutionary implications, 249
 female, 240, 241, 245, 246, 247
 primates, 248
 twins, 240
Homozygous, 84
Hormones, 42, 48, 218
Hormone replacement therapy, 181
5HT1β receptor, 144, 168–69, 213, 215, 216
5HT2a receptor, 189
5HT2c receptor, 186
Human Genome Project, 93, 194, 216, 218, 238, 251, 258, 271, 273, 278, 292
Huntington's disease, 21, 75, 153, 284, 292
Hyperphagia, 143, 183
Hypothalamus, 47, 106, 167, 183, 184, 185

Immigration Act of 1924, 288
Impulsivity, 142–47, 148, 151, 155, 175, 196, 199, 213, 219
Inbred strains, 91
Infradian clocks, 96
Insulin, 184, 193
Intelligence, 223, 284
Interneurons, 63, 125, 138
Intrascale correlation, 12
Introns, 81
In vitro fertilization, 291, 292
Ion channels, 35, 37, 64, 122, 130, 135, 140, 254, 256, 259, 268
IQ tests, 228–33, 285, 287, 293

IQ tests (continued)
  age, effect of, 231

Jennings, Herbert, 28, 32
Jet lag, 109, 112–115
Junk food, 190

κ opioid receptor, 205, 209
Kevles, Daniel, 279
Kidney stones, 204
Kinase, 122, 130, 255
Kleptomania, 143
Klinefelter's syndrome, 161
Knockout mice
  dopamine, 189
  dopamine transporter, 209
  5HT1β receptor, 168–69, 209, 213
  5HT2c receptor, 186
  maoa, 171–72
  nitric oxide synthase, 173
  norepinephrine transporter, 209
  NMDA receptor, 153
  serotonin transporter, 209
Konopka, Ronald, 103
Kung, Ching, 34

Language, influence on mental function, 233, 260
LDL (low density lipoprotein), 235
Learning, 64, 117–36, 127–288, 151
  associative, 71, 126–27
  intelligence and, 223
  mammals, 131–136
  nonassociative, 65
Leptin, 181–85, 190, 193, 194, 197
Lewontin, Richard, 264
Lewis, Ed, 126
Lithium, 112
LOD scores, 194
Long-term potentiation, 134, 152
lrn-1 gene, 72, 75
Luria, Salvatore, 125

Macronucleus, 27
Major depression, 148, 170
Male effect, 49, 55
maoa gene, 170–72, 236
Mania, 147
Manic-depressive disorder, 112, 170
Marmosets, 53
Marriage laws, 288
Mating types, 30
McClintock, Martha, 43–45, 54, 57
Mechanoreceptors, 42
Meiosis, 275
Melatonin, 109, 112, 113
Memory, 64, 117–36, 151
  in *Aplysia*, 118–124
  emotional, 131
  explicit (declarative), 119, 133–34
  implicit (nondeclarative), 119, 123, 130–31, 134, 135
  long term, 120, 122–23,

135, 152
physical changes associated with, 118, 121, 222
short term, 120, 122–23, 127
Mendel, Gregor, 28, 33, 83, 124, 226, 278, 280
Mendelian inheritance, 283, 284, 290
Menstrual cycle, synchronization, 44, 52
Mental function, 221–38, 282, 285
role of genes, 234
role of language, 233
variability in, 222, 224
Mental retardation, 172, 236
Meridia, 188
Mesolimbic region (brain), 205, 206, 212
Messenger RNA (mRNA), 81
Metabolic rate, 181, 186
Micronucleus, 28, 31
Minnesota Multiphasic Personality Inventory, 14
Minnesota Twins Study, 12–16, 230
Monera, 25
Morgan, Thomas Hunt, 124, 284
Morphine, 203, 204
Motor neurons, 63, 72
Müllerian hormone, 161
Multicellularity, 59
μ opioid receptor, 205, 206, 209, 212
Mushroom body, 125, 130
Mutation, 32, 33, 38, 93

Naloxone, 209
Naltrexone, 209
Nature vs. nurture, 4, 222, 227, 264, 270
Nazi Germany, 288, 289
Nematodes, 61
Nerve fibers, 66
Nerve impulse, 38
Nerve ring (*C. elegans*), 65, 67, 118, 126
Neurons, 66, 260
afferent, 63
destruction of, 235
environment, influence on, 222
facilitating, 121, 150
interneurons, 63, 254
motor, 63, 72, 125, 254
sensitization of, 121
sensory, 125, 221, 254
Neuropeptide Y, 70, 139, 189, 254
Neurospora crassa, 97–102, 111
conidiation oscillator, 98
Neurotensin, 139
Neurotransmitters, 67, 120, 121, 137–56, 193, 196, 200, 203, 212–13, 221, 236, 259, 264. *See also individual neurotransmitters*
Nicotine, 146
Nitric oxide, 172–73, 172n
Nitroglycerin, 172n
Nitric oxide synthase, 173–74
NMDA receptor, 152,

NMDA receptor (continued)
155, 236, 257, 292
Nociceptors, 42
Nomarski optics, 67
Norepinephrine, 139, 150, 171, 188, 189
npr-1 gene, 70, 254, 256
Nucleotides, 78, 271
Nucleus accumbens, 145, 146, 208

obese gene, 182
Obesity, 177, 199
  acquired vs. inherited, 190
  genetic elements in, 178
Obsessive-compulsive disorder, 264
odr-10 gene, 69
Olfaction, 45
Olfactory bulb, 46, 145
Oscillators, 97
Open field test, 87, 90
Opiates, 204–7
Opioids, 204–207, 212, 218
Opsins, 107
Osmoreceotors, 42
Ovulation, 53
Oxytocin, 139

Pain, 203–6
Pantophobiac mutation, 35, 253
Paramecia, 26–38, 60, 63, 64, 122, 135, 221, 253
Paraphilia, 143
Parkinson's disease, 146
Paxil (paroxetine), 149, 170
Partitioning (embryo), 8
pawn mutation, 34, 253
Paxil, 149
PDE. *See* phosphodiesterase
Pearson, Karl, 282
Peer group selection, 264
Peer pressure, 218–19
per gene, 103, 109, 111
Personality, 14, 19, 231, 233
PET (positron emission tomography) scans, 263
Phenotype, 33, 39, 76, 78, 90, 191, 197, 218, 222, 226, 275, 277
Phentermine, 187
Phenylthiocarbamide (PTC), 85
Pheromones, 43–57, 68, 108
  and aggression, 159, 163
  nursing behavior, 56
  social communication, 53
  sexual communication, 53
Phosphodiesterase, 129, 133, 236
Photoreceptors, 42, 101, 108, 111, 114
Phototherapy, 112–14
Pima Indians, 189–95
Pineal gland, 109, 113
Pituitary gland, 47
Placenta, 8
Plato, 227
Plomin, Robert, 237
Polymorphism, genetic, 83, 92
Preferential same-sex activity, primates, 248
Primates, 248, 281
Prolactin, 139
Proteins, 81, 180, 221
Protists, 25, 43, 59, 66

Protoctista, 25
Prozac (fluoxetine), 149, 170, 187, 189
ps-1 gene, 234
Psychological dependency (drugs), 208
Psychotherapy, 216
Puromanis, 142

Quantitative traits, 88, 94, 214, 242, 250, 256, 284, 290
Quantitative trait loci (QTL), 90, 194–95, 209–10, 245, 256, 257, 259, 277
Quantum mechanics, 267

Race, 228, 281, 286
Raphe nucleus, 145, 146
Reasoning skills, 224
Renfrew, John, 158
Retinal, 107
Reversion to the mean, 227, 283
Reward system, brain, 146
Rhodopsin, 107
Ribosomes, 81
Rod cells, 107
Rolipram, 133
Roundworms, 61
rutabaga mutation, 130, 133, 135, 255

Sacks, Oliver, 146
Satiety, 182, 186
Schizophrenia, 153, 170
Schrödinger, Erwin, 125
Scopolamine, 133
Seasonal affective disorder (SAD), 112
Second messenger, 130
Selective breeding, 87, 163–64, 282, 283
Senile dementia, 231
Sensation seeking, 143, 200
Sensitization, neuronal, 121
Sensory systems, 42
Serenics, 168, 170
Serotonin, 122, 134, 139, 144, 149, 167–68, 185–190, 196, 212–13, 259
Serotonin-1β receptor. *See* 5HT1β receptor
Serotonin transporter, 144, 149, 154, 168, 264
Sertraline. *See* Zoloft
Set-point weight, 179, 180
Sex, 30
Sexual preference, 239–52
Sexually indifferent stage, 160
shaker mutation, 135
Short tandem repeats (STRs), 273
Sibutramine. *See* Meridia
Sickle-cell anemia, 250, 293
Silkworms, 43
Smell, 42, 47, 68
Social behavior, 281
Soma, 9, 38, 60
Somatostatin, 139
Spearman, Charles, 285
Sponges, 60
Spores, neurospora, 97
sry gene, 163–64, 246
SSRI (selective serotonin reuptake inhibitor), 149, 151, 170
Stent, Gunther, 125

Sterilization laws, 288, 289
Stern, Wilhelm, 287
STR. *See* Short tandem repeats
Stress, 49, 53, 149, 220
Sturtevant, Albert, 126
Substance abuse, 188, 201–20
  genetic elements in, 200
Substance P, 139
Suicide, 142, 143, 148, 212
Suprachiasmatic nucleus (SCN), 106–7, 109, 111, 113
Synapse, 66, 139, 140, 147, 222
Synaptic remodeling, 118, 121, 222, 262, 263, 264

Taste, 42, 68
Testosterone, 49, 163–67, 247
Thermoreceptors, 42
Thyrotropin, 139
tim gene, 103, 210
Tolerance (to drugs), 201, 203
Tourette's syndrome, 154
Transmission genetics, 271
Transporter receptor, 140
Trichinosis, 61
Trichotillomania, 143
Tryptophan, 144, 171, 172, 186
Tryptophan hydroxylase, 172, 215
Turner's syndrome, 161
Twins, 147, 148, 258, 263
  aggression in, 164
  alcoholism, 210–11
  anorexia, 188
  biological nature of, 7–10
  BMI, 179
  dizygotic (fraternal), 4, 9
  IQ and, 230
  monozygotic (identical), 4, 7, 90
  personality, 10–19
  Siamese, 8
  substance abuse in, 200
Tyrosine, 146

Unipolar disorder, 147
Ultradian clocks, 96

Variability, genes and, 83
Vasopressin, 139, 256
Verbal skills, 224
Viagra, 172n
Viruses, 26
Vitamin A, 101, 107
Vitamin B, 101, 108
Voles, 256
Vomeronasal organ, 45–50, 56, 57, 108

Watson, James, 81
Wechsler Adult Intelligence Scale, 229, 236, 285
Werner's syndrome, 77
Wilkins, Maurice, 81
Withdrawal (from drugs), 201

X-chromosome, 160, 242, 245
X-linked traits, 242

Y-chromosome, 160, 162, 247
Yeast, 221

Zoloft (sertraline), 149, 170
Zygote, 7, 291